InspireScience

Be a Scientist Notebook

Student Journal

Grade 4

Mc
Graw
Hill
Education

mheducation.com/prek-12

STEM McGraw-Hill is committed to providing instructional materials in Science, Technology, Engineering, and Mathematics (STEM) that give all students a solid foundation, one that prepares them for college and careers in the 21st century.

Send all inquiries to:
McGraw-Hill Education
8787 Orion Place
Columbus, OH 43240

ISBN: 978-0-07-678226-0
MHID: 0-07-678226-3

Printed in the United States of America.

9 10 11 12 13 14 QSX 23 22 21 20 19

Our mission is to provide educational resources that enable students to become the problem solvers of the 21st century and inspire them to explore careers within Science, Technology, Engineering, and Mathematics (STEM) related fields.

TABLE OF CONTENTS

Energy and Motion

Transfer of Energy

Structures and Functions of Living Things

Wave Patterns and Information Transfer

HANNAH
Welder

TABLE OF CONTENTS

Patterns of Earth's Changing Features

Natural Hazards

Energy from Natural Resources

VKV Visual Kinesthetic Vocabulary

Check out the activities in every lesson!

HANNAH
Welder

Inspire Science

This is your own personal science journal where you will become scientists and engineers. Use this book to answer questions and solve real-world problems.

This is YOUR journal! Personalize it!

MALIK
Photonics Engineer

 Name ______________________ Date __________

Energy and Motion

Science in Our World

Roller coasters travel fast! They travel up and down hills. Some even make loops! Look at the photo of the roller coaster. What questions do you have?

___

___

___

___

___

___

___

___

___

Key Vocabulary

Look and listen for these words as you learn about energy and motion.

acceleration	conservation of energy	contact force
energy	energy transfer	gravity
inertia	kinetic energy	noncontact force
potential energy	speed	velocity

Name ______________________ Date ____________

STEM Career Connection

Mechanical Engineer

When you are a mechanical engineer, there is no typical day! Every day is different. We have many responsibilities. There are different types of mechanical engineers. I work on roller coasters. Some days, I use the computer at my office to make designs. Other days, I test prototypes. Still other days, I meet with scientists, other engineers, and construction crews to talk about designs and construction.

Making sure roller coasters are safe is one of the most important parts of my job.

Draw and label a diagram to show how you think roller coasters work.

Science and Engineering Practices

I will construct explanations and design solutions.

I will ask questions and define problems.

 Name ______________________ Date __________

Energy and Speed

When Does It Have Energy?

Four friends were playing kickball. They each had different ideas about the ball and energy. This is what they said:

Lily: *The ball has to be on the ground, not moving, to have energy.*

Mike: *The ball has to be moving to have energy. It doesn't matter how fast it is moving.*

Otto: *The ball has to be moving very fast to have energy.*

Ava: *The ball has energy when it is both moving and not moving.*

Who do you agree with most? ______________________

Explain why you agree.

____

____

____

____

Name ______________________ Date ____________

Science in Our World

Watch the video of the race car. What questions do you have?

____

____

____

____

____

____

Read about an automotive engineer and answer the questions on the next page.

STEM Career Connection

Automotive Engineer

We are making great progress on the design of the new solar-powered bus! Today I completed the computer model of the vehicle. It looks great!

Tomorrow I will present the design to the rest of my team. They are concerned about the speed at which the bus will be able to travel. Many older versions use too much energy and go very slowly over short distances. My new design will be able to carry people throughout the city quickly, and it will use less energy.

If my team approves the design, our next step will be to decide what type of materials we should use for the exterior and interior parts of the bus. My team and I will have to consider many factors, such as the strength, weight, and cost of the materials.

RILEY
Automotive Engineer

 Name ______________________ Date __________

1. What are the next steps if the automotive engineer's design is approved?

____

____

____

____

2. Why might it be helpful to have a team of different people working on a project instead of just one person?

____

____

____

Essential Question
How are energy and speed related?

____

____

____

____

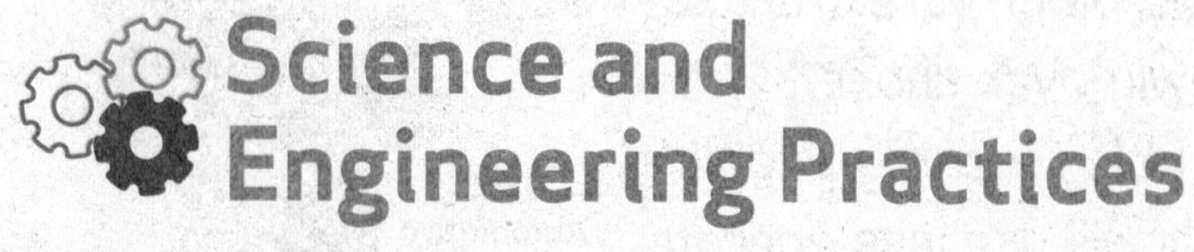

Science and Engineering Practices

I will construct explanations.

Like an engineer, you will use evidence to explain what you learn in this lesson.

Name ____________________ Date __________ EXPLORE

Inquiry Activity
The Moving Marble

How will the height of a ramp affect the speed of a marble?

Make a Prediction Which do you think will move faster: a marble rolling down a ramp from a low height or a marble rolling down a ramp from a higher height? Explain your prediction.

____

____

____

____

Materials

- ☐ 4 books
- ☐ cardboard tube
- ☐ tape
- ☐ stopwatch
- ☐ marble

Carry Out an Investigation

1. Stack three books on top of each other. Place one end of a cardboard tube on top of the stack. The other end of the tube will touch the spine of the fourth book. Tape the tube in place.
2. Start the stopwatch when you roll the marble down the tube. Stop the timer when you hear the marble hit the fourth book.
3. **Record Data** Record the time in the data table. Repeat steps 2 and 3 two more times.
4. Repeat steps 2 and 3 three times with two books stacked. Then repeat steps 2 and 3 three times with only one book. Calculate the average time for each trial.
5. Make a bar graph on a separate sheet of paper comparing the results for three books, two books, and one book.
6. **Analyze Data** Circle the stack of books with the fastest times.

	Marble Travel Time (seconds)			
	Trial 1	Trial 2	Trial 3	Average
Three Books				
Two Books				
One Book				

 Name ______________________ Date __________

Communicate Information

1. Did your results match your prediction? Explain.

2. In which trial do you think the marble had the most energy? Explain.

3. **Construct an Explanation** What happened to the speed of the marble when you used fewer books? Why do you think this happened?

Glue your graph here.

Name ______________________ Date __________ EXPLAIN

Obtain and Communicate Information

Vocabulary

Use these words when explaining speed and energy.

speed	velocity	acceleration
energy	potential energy	kinetic energy

Measuring Motion

Read pages 278–280 in the ***Science Handbook.*** Answer the following questions after you have finished reading.

1. How do you calculate speed?

____

____

2. Describe how velocity is different from speed.

____

____

3. How are speed and acceleration different?

____

____

____

Energy

Read pages 296–299 in the ***Science Handbook.*** Answer the following questions after you have finished reading.

4. What happens to the potential energy of an object when the object is raised higher?

____

____

5. Explain how kinetic energy and speed are related.

____

____

____

Speed and Energy

Watch the video *Speed and Energy.* Answer the question after you have finished watching.

6. Use one of the examples from the video to describe the relationship between energy and speed.

____

____

____

Energy, Mass, and Speed

Read pages 282–283 in the ***Science Handbook.*** Answer the following questions after you have finished reading.

7. What is a force?

____

____

8. How does the size of a force affect the acceleration of an object?

____

____

9. How does the mass of an object affect the acceleration of an object when you apply a force?

____

____

____

FOLDABLES®

Cut out the Notebook Foldables tabs given to you by your teacher. Glue the anchor tabs as shown below. Use what you have learned to describe how the soccer ball and the apple could have kinetic or potential energy.

Glue anchor tab here.

Glue anchor tab here.

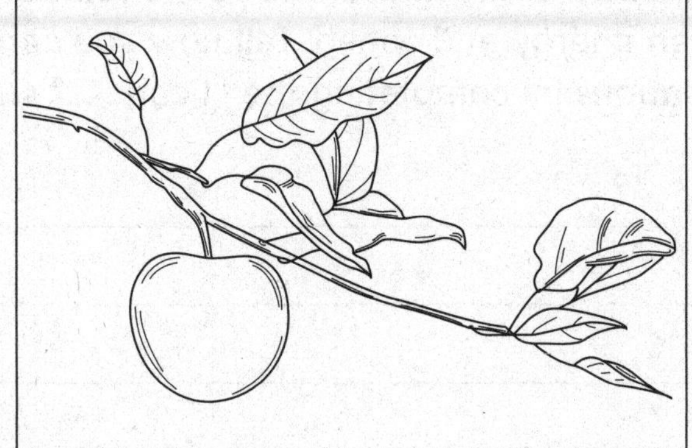

 Name ______________________ Date __________

The Moving Marble

10. Revisit *The Moving Marble* activity on page 7. Draw and label a diagram of your setup. Use lesson vocabulary to label the type of energy that was present throughout the marble's movements.

11. How does your diagram show the relationship between potential and kinetic energy? Use lesson vocabulary in your response.

Science and Engineering Practices

Use examples from the lesson to explain what you can do!

Think about the evidence that you have collected throughout the lesson that shows the relationship between energy and speed. Tell how you can construct explanations by completing the "I can . . ." statement below.

I can ______________________

Research, Investigate, and Communicate

Inquiry Activity

Mass Matters

Materials

- safety goggles
- 2 books
- thin, flat board
- meterstick
- masking tape
- 500-mL plastic bottle with screw cap
- pan balance
- plastic cup
- water
- graduated cylinder

You will observe how mass affects the kinetic energy of an object. You will use objects with different masses traveling at the same speed.

Write a Hypothesis Read the activity. How will increasing the mass of the water bottle affect the distance the plastic cup moves? Write a hypothesis in the form of an "If..., then..." statement.

____

____

____

Carry Out an Investigation

BE CAREFUL Wear safety goggles to protect your eyes.

1. Stack two books. Place one end of the board on the books to form a ramp. Mark the ramp where the ramp touches the books to use as the starting point for the bottle.
2. Measure 20 cm from the bottom of the ramp and mark the distance with tape. This will be the starting point of the cup for each trial.
3. Fill the bottle with 100 mL of water. Screw the cap on tightly. Measure the mass of the water and bottle on the pan balance, and record the measurement below.

Mass of 100-mL Water Bottle	Mass of 200-mL Water Bottle	Mass of 300-mL Water Bottle

 Name ____________________ Date ____________

4. Place the empty cup at its starting point. Place the bottle at its starting point. Let go of the bottle and allow it to run into the cup and move it.

5. **Record Data** When both the cup and the bottle have stopped moving, use the meterstick to measure the distance the cup moved from its starting point. Record your data.

6. Repeat steps 4 and 5 two more times. Record your data.

7. Add 100 mL more water to the bottle. Measure the mass. Repeat steps 4-6 to get three trials with the new mass. Record your data.

8. Add 100 mL more water to the bottle. Measure the mass. Repeat steps 4-6 to get three trials with the new mass. Record your data.

9. **Analyze Data** Using the data collected, create a line graph on a separate sheet of paper showing the relationship between the mass of the bottle and the average distance the cup moved. Use a ruler if necessary to create straight lines. Label the X and Y axes and title the graph.

	Distance Cup Moved with 100-mL Water Bottle	Distance Cup Moved with 200-mL Water Bottle	Distance Cup Moved with 300-mL Water Bottle
Trial 1			
Trial 2			
Trial 3			
Average Distance			

Glue your graph here.

Communicate Information

1. What evidence did you observe in your investigation to explain the relationship between kinetic energy and mass?

2. Why was it important that the height of the ramp did not change?

3. **Construct an Explanation** Does the data support your prediction? Explain.

 Name ____________________ Date __________

Performance Task
Test Toy Cars

Be an automotive engineer and perform tests on the toy car. You will investigate what happens when the amount of energy applied to the car changes.

Materials

- ☐ safety goggles
- ☐ masking tape
- ☐ meterstick
- ☐ 2 wooden blocks with securely fastened nails
- ☐ rubber bands
- ☐ toy car
- ☐ stopwatch

Make a Prediction You will use rubber bands to generate different amounts of energy. Predict which number of rubber bands will make the car move fastest, second fastest, and slowest.

	Prediction
One Rubber Band	
Two Rubber Bands	
Three Rubber Bands	

Carry Out an Investigation

BE CAREFUL Wear safety goggles to protect your eyes.

1. Place a 12-inch strip of masking tape on a tile floor. This is the "starting line." Hold the blocks with nails pushed in them on opposite ends of the tape. Stretch a rubber band between the nails.
2. Using masking tape, place a "finish line" on the floor 1.5 meters from the rubber band.
3. Pull a toy car back against the rubber band a few centimeters. Mark this starting point with a piece of tape and use it for the rest of the investigation.
4. **Record Data** Let go of the car. Have a partner use a stopwatch to time how long it takes for the car to cross the finish line. Record the data in the table.
5. Repeat this for two more runs. Then calculate the average time it took the car to reach the finish line.
6. Repeat steps 3 through 5 with 2 rubber bands and then with 3 rubber bands.

Use evidence from an experiment to explain the relationship between speed and energy.

	First Run Time	Second Run Time	Third Run Time	Average Time
One Rubber Band				
Two Rubber Bands				
Three Rubber Bands				

7 **Analyze Data** Using the data collected, create a line graph showing the relationship between the number of rubber bands used and the average time needed for the car to travel 1.5 meters. Use a ruler if necessary to create straight lines. Label the X and Y axes and title the graph.

Communicate Information

1. **Construct an Explanation** Do the data support your prediction? Why or why not?

____

____

____

____

____

____

____

____

Glue your graph here.

Crosscutting Concepts

Patterns

2. Did you notice a pattern in the data collected?

3. Use the evidence that you collected to explain the relationship between the speed and energy of the car.

4. Infer what would happen if you used four rubber bands.

Name ______________________ Date __________

Essential Question

How are energy and speed related?

Think about the video of the race car at the beginning of the lesson. Explain how the energy and speed of the race car are related.

____

____

____

____

____

Science and Engineering Practices

Review the "I can . . ." statement you wrote arlier in the lesson. Explain what you have accomplished in this lesson by completing the "I did . . ." statement.

I did ____

____

____

____

____

____

____

____

____

Energy Change in Collisions

Toy Car Crash

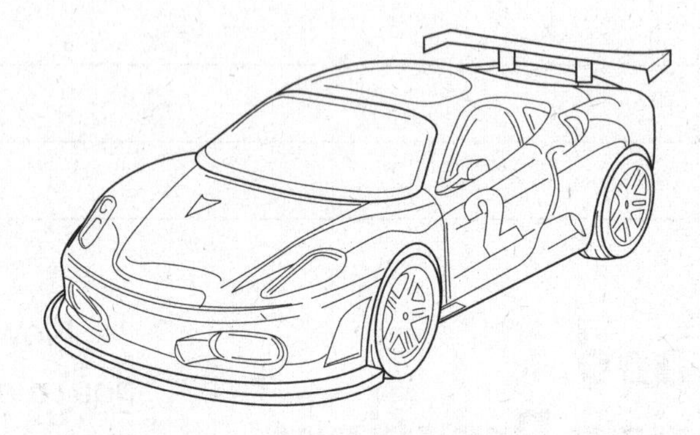

Three friends were playing with their toy cars. Bert crashed his car into Alonso's car. They each had different ideas about what was transferred during the crash. This is what they said:

Bert: *I think the force from my car was transferred to your car.*

Alonso: *I think the energy from your car was transferred to my car.*

Gus: *I think both the force and energy from Bert's car was transferred to Alonso's car.*

Who do you agree with the most? ________________

Explain why you agree.

____

____

____

____

____

____

Name ______ Date ______

Science in Our World

Watch the video of the car crash. What questions do you have?

______

______

______

______

______

______

Read about a biomechanical engineer and answer the questions on the next page.

STEM Career Connection

Biomechanical Engineer

I just completed the design of my first product while working as a biomechanical engineer at a sporting goods company. I designed a safer football helmet. It took a lot of research, many tests, and revisions.

We tested the different designs by placing computers in the helmets to measure the impact of collisions. The new design protects players from head injuries that can occur during collisions. Our helmet reduced the impact of collisions more than any other design on the market!

Name ______________________ Date __________

1. How do biomechancial engineers help keep football players safe during collisions?

2. How did this biomechanical engineer test her designs?

Essential Question
What happens when objects collide?

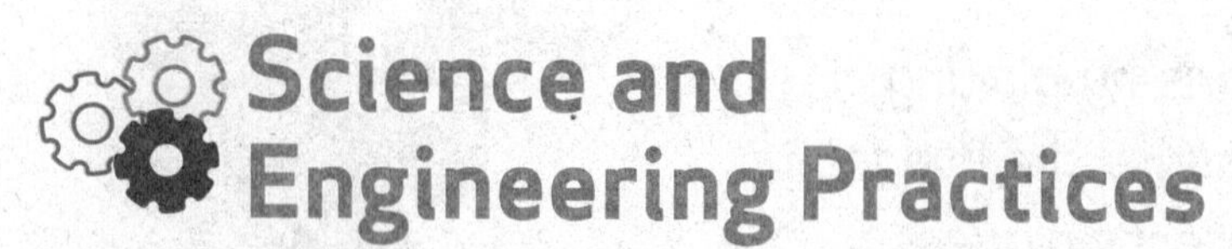

Science and Engineering Practices

I will ask questions and define problems.

Like the biomechanical engineer, you will ask questions and define problems in this lesson.

Name ______________________ Date ____________ EXPLORE

Inquiry Activity
Collision Variables

Materials
- ☐ 3 different types of balls
- ☐ meterstick

How will different types of balls react when they collide with a hard surface?

Make a Prediction Read the activity. What do you think will happen with each ball when you drop it onto a hard surface?

Type of Ball	Prediction

Carry Out an Investigation

1. Hold the meterstick straight up with the "0" end on a hard surface.
2. Hold the first ball next to the 55 cm mark on the meterstick. Drop the ball onto the hard surface.
3. **Record Data** Watch what happens to the ball when it hits the surface. If it bounces, record the height that it bounced in the table. If the ball does not bounce, record 0 cm.

 Name ______________________ Date __________

4 Repeat steps 1–3 for each ball. Continue repeating until you have collected data from three trials.

Type of Ball	Ball Bounce-back Height (cm)			
	Trial 1	Trial 2	Trial 3	Average

5 **Analyze Data** Determine the average bounce-back height for each ball. Circle the one with the highest bounce-back height.

Communicate Information

1. **Infer** Why do you think the ball you circled had the highest bounce-back height?

____

____

2. What do you think would happen if you dropped the balls from a higher position on the meterstick?

____

____

Crosscutting Concepts

Energy and Matter

3. Do you think the ball had energy before it was dropped, when it was moving, or in both cases?

____

____

____

____

Name ______________________ Date __________

Obtain and Communicate Information

Vocabulary

Use these words when explaining energy transfer in collisions.

contact force	noncontact force	inertia
energy transfer	conservation of energy	gravity

Forces

Read page 282 in the ***Science Handbook***. Answer the following question after you have finished reading.

1. What is the difference between contact forces and noncontact forces?

____

____

____

Gravity and Friction

Read pages 286–287 in the ***Science Handbook***. Answer the following questions after you have finished reading.

2. What effect does Earth's gravity have on objects on Earth? Give an example of this effect.

____

____

____

3. What is friction?

____

____

____

 Name ______________________ Date __________

Energy Changes

Read pages 302–303 in the ***Science Handbook***. Answer the following questions after you have finished reading.

4. What does transforming energy mean?

____

____

5. What does transferring energy mean?

____

____

____

6. Describe the transformation of energy between kinetic and potential energies using a roller coaster as an example.

____

____

____

____

____

Conservation of Energy

Explore the digital interactive *Conservation of Energy* about energy transfer in a car crash. Answer the question after you have finished.

7. What happens to the kinetic energy from the moving cars when they collide?

____

____

____

Name ______________________ Date __________ EXPLAIN

Inertia and Momentum

Read *Inertia and Momentum* about how forces act on objects. Answer the following questions after you have finished reading.

8. Use the word *inertia* to explain why you feel like the seat is pushed into your back when a vehicle starts moving.

____

____

____

9. Why is it important to use a seat belt when you are in a vehicle?

____

____

____

10. Which has the greater momentum, a horse with a mass of 450 kg running with a velocity of 13 m/s or a horse with a mass of 450 kg running with a velocity of 4 m/s? Explain your answer.

____

____

____

____

____

Science and Engineering Practices

Use examples from the lesson to explain what you can do!

Think about the investigation you conducted. Tell how you can ask questions or define problems by completing the "I can . . ." statement below.

I can ____

____

____

 Name ______________________ Date ____________

FOLDABLES®

Cut out the Notebook Foldables tabs given to you by your teacher. Glue the anchor tabs as shown below. Use the vocabulary words you have learned to describe what is happening in the picture.

Glue anchor tab here.

Name ______________________ Date __________ ELABORATE

Research, Investigate, and Communicate

Newton's Cradle

Investigate energy transfer in collisions by exploring the simulation. Answer the questions after you have finished.

1. What happens to the potential energy of a sphere when you lift it higher or lower?

____

____

____

2. Ask a question about what will happen if you lift more than one sphere. Be specific about the number of spheres. Use the simulation to answer the question.

____

____

____

____

____

____

____

3. Draw and label a diagram that shows how energy is transferred in a Newton's Cradle device.

 Name ______________________ Date __________

Performance Task
Protect an Egg

As a biomechanical engineer, you design solutions that protect people. Biomechanical engineers who work on cars use crash test dummies to test protective designs. You will use an egg as your crash test subject. You need to design a solution that prevents the egg from breaking when dropped from a height of two meters.

Materials

- safety goggles
- raw egg
- small box
- variety of packing materials
- measuring tape or meterstick
- masking tape
- garbage bag

Define a Problem Describe the problem you will be solving in your own words.

____

____

____

Design a Solution

BE CAREFUL Wear safety goggles. Use caution when climbing the ladder.

1. Using the box and packing materials, design a solution that will protect your egg when it is dropped from a height of 2 meters.
2. Make a sketch of your design below. Build a model of the design based on your sketch.

3 Set the ladder next to a wall. Measure a spot on the wall 2 meters from the ground. Mark the spot with masking tape. Lay out a garbage bag on the ground beneath your mark.

4 **Test** Place your egg inside your model. Drop the model from the marked height. Examine what happened to the egg.

5 Evaluate your design. Make changes to the sketch and rebuild as needed.

6 Test your new design.

Communicate Information

1. Why did you choose to use the materials that you did?

____

____

2. Explain the energy changes that occurred as your model and the egg collided with the floor.

____

____

____

____

____

3. How do you think the process you followed is similar to what a biomechanical engineer might do?

____

____

____

____

EVALUATE Name ____________________ Date __________

Essential Question

What happens when objects collide?

Think about the video of the crash test dummy you watched at the beginning of the lesson. Explain what is happening to the energy during the collision.

____

____

____

____

Science and Engineering Practices

Now that you're done with the lesson, share what you did!

Review the "I can . . ." statement you wrote earlier in the lesson. Explain what you have accomplished in this lesson by completing the "I did . . ." statement.

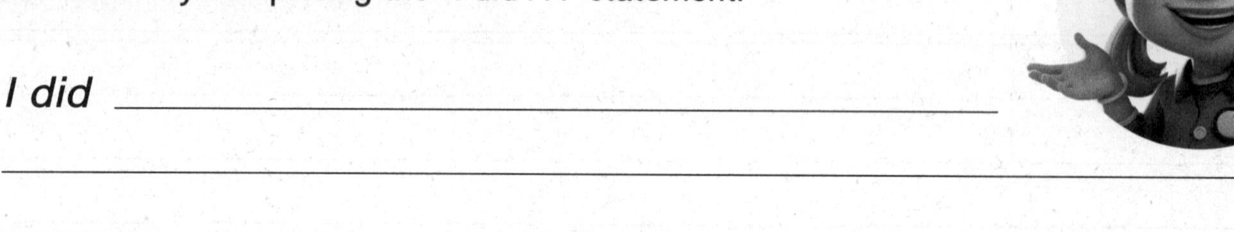

I did ______________________________________

____

____

____

____

____

Energy and Motion

Performance Project
Design a Roller Coaster

Materials

- [ ] 2 marbles
- [ ] plastic cup
- [ ] cardboard tubes
- [ ] tape

Define a Problem Use what you have learned about energy, motion, mass, and collisions to design a roller coaster that can knock a marble into a cup. You need to make sure that the marbles have the amount of energy that makes the second marble fall into the plastic cup while the first marble stays on the track. Write this problem in your own words.

____

____

____

____

____

____

Design a Solution

1. On the next page, draw a sketch of the roller coaster design you intend to use to solve the problem.
2. Make a model from your sketch.
3. **Test** Place a marble at the end of the roller coaster above the cup. Place a second marble at the beginning of the roller coaster.

 Name ______________________ Date ____________

Energy and Motion

4 Release the marble and allow it to roll down the first hill. Watch what happens to the marbles at the end of the roller coaster.

5 **Analyze Data** What happened during your test run? Explain what your data revealed.

___

___

___

6 Adjust the design of your roller coaster until you are able to get only the second marble to land in the cup.

Name ______________________ Date ____________ MODULE WRAP-UP

Roller coasters are designed using knowledge of the relationship between energy and speed.

Explain the relationship between energy and motion that you observed in the roller coaster model. Use the words *kinetic energy* and *potential energy* to describe the motion of the marble. Use the word *transfer* to explain the collision between the two marbles.

Explore More in Our World

Did you learn the answers to all of your questions from the beginning of the module? If not, how could you design an experiment or conduct research to help answer them?

 Name ______________________ Date __________

Transfer of Energy

Science in Our World

Look at the photo. Have you ever seen these machines? They can be found in fields, along mountaintops, or even offshore in the ocean. These machines are called wind turbines, and they help harness the energy in wind! What questions do you have about wind turbines and how they help up get energy?

____

____

____

____

____

____

Key Vocabulary

Look and listen for these words as you learn about transfer of energy.

circuit	conduction	constraint
convection	criteria	design process
electromagnetic spectrum	generator	photon
radiation	solar cell	thermal energy

Name ______________________ Date ____________ MODULE OPENER

How can I use what I know about energy transfer to design a device that generates electricity?

FINN
Construction Manager

STEM Career Connection

Wind Energy Engineer

Field Notes

Date: April 4, 2015

Tower #123

Wind Speed: 16 kilometers per hour

Wind Output: 7,202 megawatts

Tower Height: 85 meters

Rotor Width: 100 meters

Draw and label a diagram to show how you think wind turbines are used to transfer energy.

Science and Engineering Practices

I will plan and carry out investigations.

I will construct explanations and design solutions.

 Name ____________________ Date __________

Types of Energy Transfer

Movement of Energy

Energy can move from place to place. Put an X in all of the boxes that are examples of energy moving from one place to another.

Thunder claps loudly	A lightbulb lights up	A car crashes into a wall
Ice cubes stay frozen in the freezer	An ice cube melts in the hot sun	An electric fan turns
Hot water cools off	Wood burns in a fireplace	A book sits on a shelf
A car horn beeps	A baseball bat hits a ball	A bowing ball knocks over pins

Explain your thinking. What kinds of evidence show that energy moves from one place to another?

____

____

____

____

____

Science in Our World

Look at the photo of a person being launched from a cannon. What questions do you have?

____

____

____

____

____

____

Read about a physicist and answer the questions on the next page.

STEM Career Connection

Physicist

When I was little, I was one of those kids who was always asking "Why?" That is part of the reason why I began studying physics. It explains what happens in the world around us. If a question has not been answered, science gives us the tools to try to answer it. I perform investigations to see how changing one variable affects another. In my research, I am trying to help find renewable energy solutions. With the help of a team of scientists and engineers, I perform experiments with different alternative energy sources. We test how changing different variables affects the amount of electricity that can be generated.

Physicists study a wide range of topics. Physicists have to understand how energy transfers from place to place to be able to answer questions and solve problems.

MALIK
Photonics Engineer

Name ______________________ Date __________

1. Why did this person become a physicist?

____

____

____

____

2. What is an important part of an investigation that this physicist mentions?

____

____

____

Essential Question

How is energy transferred?

____

____

____

____

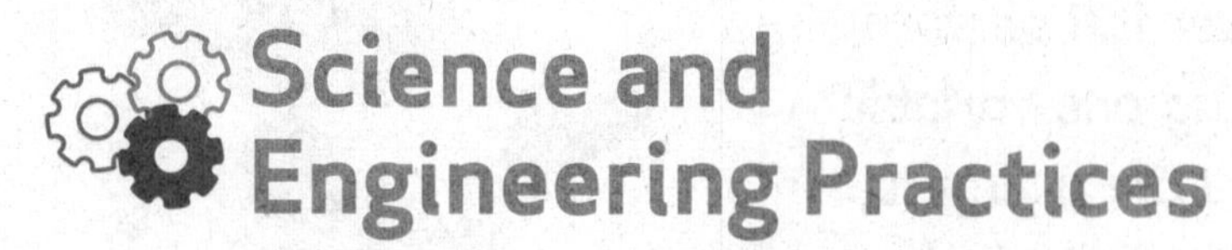

Science and Engineering Practices

I will plan and carry out investigations.

In this lesson, you will plan and carry out investigations just like a physicist.

Name ______________________ Date __________

Inquiry Activity

Demonstration of Energy Transfers

How does energy move from place to place? You will make observations as you look for evidence that energy is moving.

Materials

- [ ] radio
- [ ] bowl of water
- [ ] beaker of water
- [ ] hot plate
- [ ] cutout spiral-shaped paper on string

Make a Prediction Observe the setup for each of the demonstrations. What do you think might happen in each setup? Explain your prediction.

____

____

____

____

____

Carry Out an Investigation

BE CAREFUL: Do not touch any of the materials in this investigation.

1. At the first station, observe the radio on a desk. Observe its effect on the bowl of water.
2. **Record Data** Record your observations as the radio is turned on.

____

____

____

____

3. At the second station, observe, but do not touch, the beaker of water sitting on top of a hot plate. Observe its effect on the paper spiral.
4. **Record Data** Record your observations as the paper spiral on a string is held above the beaker of water.

____

____

____

Name ______________________ Date __________

Communicate Information

1. How does your prediction compare to what you observed at the first station?

2. What do you think caused the water to move at the first station?

3. How does your prediction compare to what you observed at the second station?

4. What do you think caused the paper to move at the second station?

Name ______________________ Date ____________ EXPLAIN

Obtain and Communicate Information

Vocabulary

Use these words when explaining types of energy transfers.

thermal energy	sound energy	heat
conduction	convection	radiation
conductor	insulator	

Other Forms of Energy

Read about different types of energy on pages 300–301 in the ***Science Handbook***. Answer the questions after you have finished reading.

1. Draw a circle around the forms of energy that are types of kinetic energy. Draw a rectangle around the forms of energy that are types of potential energy.

chemical electrical thermal nuclear sound

2. Use clues from the text to decide whether light energy would be considered kinetic energy or potential energy. Explain your answer.

____

____

____

____

____

 Name ______ Date ______

Energy Transfers

Watch *Energy Transfers* on how energy is moved from place to place. Answer the questions after you have finished watching.

3. How is energy transferred during a roller coaster ride?

4. What type of energy transfer causes things, like ice cream, to melt?

Energy Transfers in the Classroom

Explore the Digital Interactive *Energy Transfers in the Classroom* on common energy transfers that occur in a classroom. Answer the question after you have finished.

5. List three of the energy transfers you learned about. Give another example of each type of energy transfer you listed. Explain how energy is transferred in each example.

Heat Energy

Read page 304 in the ***Science Handbook***. Answer the questions after you have finished reading.

6. Circle which has more thermal energy: a cup of boiling water or a cup of warm water. Explain why you chose that answer.

7. If you place an ice cube in a glass of room-temperature water, which way does the heat flow? How do you know?

Heat Transfer

Read pages 308–309 in the ***Science Handbook***. Answer the question after you have finished reading.

8. Use drawings to illustrate examples of conduction, convection, and radiation.

Name ______________________ Date __________

FOLDABLES

Cut out the Notebook Foldables tabs given to you by your teacher. Glue the anchor tabs as shown below. Use what you have learned to define each word using the picture.

Glue anchor tab here.

Thermal Conductivity

Read page 310 in the ***Science Handbook***. Answer the following questions after you have finished reading.

9. Explain what a thermal conductor is.

10. What is a thermal insulator?

11. Which allows energy transfers to happen more easily, thermal insulators or thermal conductors?

Science and Engineering Practices

Use examples from the lesson to explain what you can do!

Think about the evidence you have collected throughout the lesson that shows different types of energy transfer. Tell how you can carry out investigations to produce evidence to be used for explaining energy transfers by completing the "I can . . ." statement below.

I can _______________

Research, Investigate, and Communicate

Energy Transfer Through Matter

Investigate how heat, sound, and light energy transfer through matter by conducting the simulation. Answer the questions after you have finished.

1. Heat—Use what you have observed in the simulation to explain how energy moves through a solid when heated.

2. Sound Waves—Use what you have observed in the simulation to explain how sound energy is transferred to your ear.

3. Light—Use what you have observed in the simulation to explain how a transfer of energy is demonstrated with light.

Performance Task
Energy Transfer Machine

You have learned about different types of energy transfer. Now, it is time to show what you have learned by completing an investigation. As a physicist, you will plan and carry out an investigation to design a Rube Goldberg device that demonstrates energy transfers.

Rueben "Rube" Goldberg was an engineer who used his knowledge of how things work to draw fun cartoons. He is best known for his "inventions." A Rube Goldberg device is an elaborate setup with many different parts, such as arms, wheels, gears, handles, ramps, string, pulleys, and cups. The parts of the device are put into motion by balls, pails, boots, balloons, and even live animals!

Using Rube Goldberg's drawings as a model, you will design your own device for solving a problem or completing a task. You will use the following steps to complete your design:

1. Research Rube Goldberg devices.
2. Choose a problem or task that your device will solve or complete.
3. Use the research you completed to plan, design, and draw your device. Include labels for the energy transfers that occur in your device.

 Name ________________ Date ________

1. **Research** Using resources provided by your teacher, research Rube Goldberg devices. Take notes about the different types of energy transfers that occur in Rube's contraptions. List parts of the devices that you would like to include in your design. Ask your teacher for more paper, if needed.

2. What problem or task will your device solve?

3. **Make a Model** Draw your device. Make sure your device includes at least three energy transfers. Label the name of each part as well as the energy transfers in your diagram. Give your device a name.

Crosscutting Concepts
Energy and Matter

4. Explain the different ways your device shows energy being transferred between objects.

____

____

Essential Question
How is energy transferred?

Think about the photo of a person being launched from a cannon at the beginning of the lesson. Explain how energy is being transferred from the cannon to the person.

____

____

____

____

Science and Engineering Practices

Now that you're done with the lesson, share what you did!

Review the "I can . . ." statement you wrote earlier in the lesson. Explain what you have accomplished in this lesson by completing the "I did . . ." statement.

I did ______________________________

____

____

____

____

Name ______________________ Date ____________

Transfer of Energy by Electricity

Energy Transfer and Electric Current

Four friends were talking about energy transfer and electric current. They each had different ideas. This is what they said:

Jack: *I think energy from a moving object can be transferred by electric current.*

Jumon: *Jack, you are wrong. It's the other way around. I think energy from electric current can be transferred to a moving object.*

Ida: *You are both right. Energy can be transferred either way.*

Mayumi: *All of you are wrong. Energy is not transferred by electric current or moving objects.*

Who do you agree with the most? ______________________

Explain why you agree.

____

____

____

____

____

Name ________________________ Date ____________

Science in Our World

Look at the photo of the flashlight. What questions do you have?

____

____

____

____

____

____

Read about an electronics engineer and answer the questions on the next page.

STEM Career Connection

Electronics Engineer

Today was a long day! We spent many hours trying to fix problems that we are seeing in the new device we are working on. The device is a new type of robot vacuum. We designed the electrical system using a computer. Then we built a working model. But we have a long way to go before this product is ready to be sold. There will be many rounds of testing and improvement to make sure the robot vacuum works as well as we want it to.

MALIK
Photonics Engineer

Name ____________________ Date __________

1. Why do you think an electronics engineer must test a design many times?

2. What do you find interesting about an electronics engineer's career?

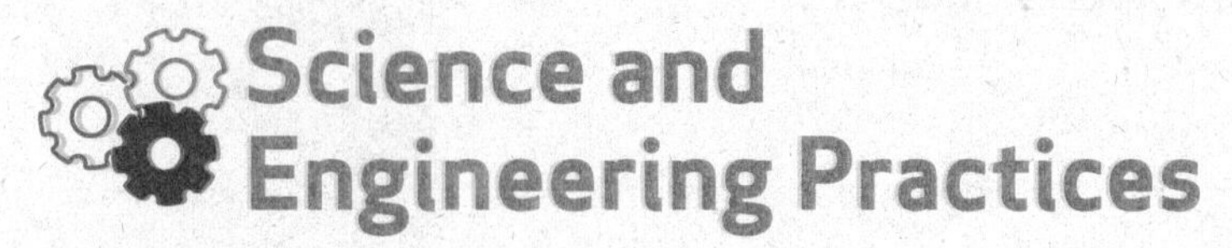

Essential Question

How do electric currents transfer energy?

Science and Engineering Practices

I will plan and carry out investigations.

In this lesson you will use evidence to help explain what you have learned, just like an electronics engineer!

Inquiry Activity
Simple Electricity

Materials

- [ ] 2 flashlight batteries
- [ ] 3 pieces of wire
- [ ] switch
- [ ] lightbulb
- [ ] scissors

How can you produce light with the supplied materials? You will investigate how to make a lightbulb light up.

Make a Prediction What effect will using more batteries have on a lightbulb?

____

____

____

____

Carry Out an Investigation

1. With your partner, determine how to use one of the batteries and the rest of the materials provided to light the bulb.
2. Draw a diagram showing how you were able to light the bulb.

3 Use the scissors to cut one of the wires in your device in the middle. Add the second battery to the device where the wires have been cut. How did this affect the lightbulb?

Communicate Information

1. How did the results of adding another battery compare with your prediction? Explain.

2. Where is the energy coming from to make your device work?

3. What types of energy transfers have you observed in this activity?

Name ______________________ Date __________ EXPLAIN

Obtain and Communicate Information

Vocabulary

Use these words when explaining energy transfers by electricity.

electric current	circuit	switch
resistor	electromagnet	generator

Uses of Electricity

Watch *Uses of Electricity* on how people use electricity every day. Answer the questions after you have finished watching.

1. List what electricity is used to produce.

____

____

2. In what ways do you use electricity every day?

____

____

Electric Current

Read pages 314–315 in the ***Science Handbook***. Answer the questions after you have finished reading.

3. How is the flow of electric current similar to the flow of water in a hose?

____

____

____

____

____

____

4. What happens to energy in a resistor?

____

____

____

____

Electric Current in a Circuit

Explore the Digital Interactive *Electric Currents in a Circuit* on how energy flows in a circuit. Answer the questions after you have finished.

5. In the table, describe what happens at these points in the circuit.

Circuit Part	Function
Power source	
Switch	
Electric current	
Fan	

Name ______________________ Date ____________ EXPLAIN

FOLDABLES®

Cut out the Notebook Foldables tabs given to you by your teacher. Glue the anchor tabs as shown below. Use what you have learned to explain how an electrical current can produce light, heat, sound, and motion in this home.

Glue anchor tab here.

 Name ______________________ Date __________

Energy Sequence

Explore the Digital Interactive *Energy Sequence* on how electricity travels from a source to your home. Answer the question after you have finished.

6. Explain how electrical energy moves from its source to the point at which it can be used.

____

____

____

____

____

____

____

____

Science and Engineering Practices

Use the examples from the lesson to explain what you can do!

Think about how you carried out an investigation and made observations throughout this lesson that show how electrical energies transfer. Tell how you can plan and carry out investigations by completing the "I can . . ." statement below.

I can ______________________________

____

____

____

Research, Investigate, and Communicate

Electromagnets

Read pages 324–327 in the ***Science Handbook***. Answer the questions after you have finished reading.

1. To have a working electromagnet, what has to happen?

____

____

____

____

2. What is the difference between a motor and a generator?

____

____

____

____

How a Generator Works

Watch *How a Generator Works*. Answer the question after you have finished watching.

3. Describe the energy transfer that takes place in an electric generator. What causes this transfer to happen?

____

____

____

____

____

____

____

 Name ______________________ Date __________

Performance Task
Make It Work

Materials
- ☐ flashlight
- ☐ batteries

Look back at the questions you wrote about the flashlight at the beginning of the lesson. Now that you know about energy transfer by electricity, you can probably answer many of those questions.

As an electronics engineer, you will carry out an investigation of a flashlight to determine how it works.

Write a Hypothesis Does the position of the batteries in the flashlight have an effect on how the flashlight works?

____

____

____

____

Carry Out an Investigation

1. Carefully take apart the flashlight and observe each part to determine its function. Draw a diagram of each part and label its function.

Name ______________________ Date __________ EVALUATE

2 Put the flashlight back together. Make sure the batteries are both facing in the same direction. The raised end of each battery should be pointed towards where the lightbulb will be located. Turn the flashlight on.

3 Remove the batteries and replace them so that the batteries are backwards. The raised ends of the batteries should now be facing away from the lightbulb. Reassemble the flashlight and turn it on.

4 Remove the batteries and replace them so that the raised ends of the batteries are touching each other. Reassemble the flashlight and turn it on.

5 **Record Data** Describe what happened to the lightbulb in each step above.

Battery Arrangement	Observations
Both raised ends point toward lightbulb	
Both raised ends point away from lightbulb	
Raised ends touch each other	

Communicate Information

1. Use the evidence you have gathered to create a diagram that shows how the working flashlight is an example of an electrical circuit. Label the flow of electric current in your diagram.

2. Describe the energy transfer that occurs within the battery that allows the lightbulb to give off light.

Crosscutting Concepts

Energy and Matter

3. How is energy transferred by the electric current? Use your diagram and the evidence that you collected to help with your explanation.

4. Give evidence that electric currents transfer energy from place to place throughout your home. Give an example of how electrical currents are used to produce motion, sound, heat, and light.

Essential Question

How do electric currents transfer energy?

Think about the photo of the flashlight at the beginning of the lesson. Explain how electric currents transfer energy in electric circuits, like the flashlight you observed.

____

____

____

____

____

Science and Engineering Practices

Now that you are done with the lesson, share what you did!

Review the "I can . . ." statement you wrote earlier in the lesson. Explain what you have accomplished in this lesson by completing the "I did . . ." statement.

I did ____

____

____

____

____

Name ______________________ Date __________

Transfer of Energy by Light

Energy from Sunlight

Two friends were arguing about energy from the Sun. They each had a different idea about using the Sun's energy. Here is what they said:

Violet: *The Sun's energy is only useful when the Sun is shining. When there is no sunlight, you can't use the Sun's energy.*

Liam: *The Sun's energy can be captured and used later. The Sun does not have to be shining in order to use the Sun's energy.*

Who do you agree with the most? ______________________

Explain why you agree.

____

____

____

____

Name ______________________ Date ____________

Science in Our World

Look at the photo of the Sun shining on a snow-covered landscape. What questions do you have?

____

____

____

____

Read about a photonics engineer and answer the questions on the next page.

STEM Career Connection

Photonics Engineer

We have just started an exciting new project testing the ability of lasers to perform a new type of heart surgery. Our job will be to design a new laser tool to perform the surgery. Lasers are already used in many types of surgery, so we will start our research by looking at what is currently being used. Then we will gather data about the new surgery. We will use the data to see how we can make a new tool that will meet specific needs in the new procedure.

MALIK
Photonics Engineer

Name ______________________ Date __________

1. How will the photonics engineer know whether the design of the new laser tool is successful?

____

____

____

____

2. Where have you seen lasers being used before?

____

____

____

Essential Question

How does light transfer energy?

____

____

____

____

____

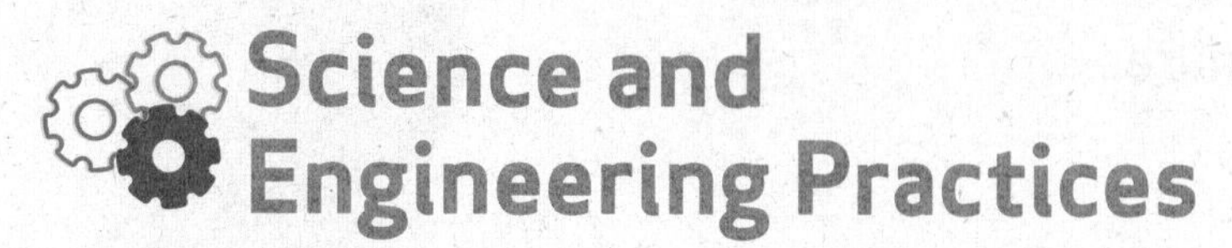

Science and Engineering Practices

I will plan and carry out investigations.

Like a photonics engineer, you will make observations, gather evidence, and explain what you learn in this lesson as you plan and carry out investigations.

Name ______________________ Date __________ EXPLORE

Inquiry Activity
Solar Circuit

Materials
- ☐ solar cell
- ☐ wires with clips
- ☐ small lightbulb
- ☐ electric buzzer
- ☐ electric motor

Can you use another source of energy other than batteries to make electric devices work? You will explore making a simple circuit using an energy source other than batteries.

Make a Prediction What do you think will happen when you expose the power source (solar cell) to different amounts of light?

____

____

____

____

____

____

Carry Out an Investigation

1. Begin by setting your solar cell in a location with very bright light.
2. Connect each of the lead wires from the solar cell to the connectors on the lightbulb.
3. **Record Data** Record your observations of the lightbulb.

____

____

____

____

4. Disconnect the lightbulb from the circuit. Then connect the lead wires from the solar cell to the electric buzzer.
5. **Record Data** Record your observations of the buzzer.

____

____

____

____

6 Repeat the same process to connect the electric motor to the solar cell.

7 **Record Data** Record your observations of the motor.

8 With one of the devices connected to the solar cell, block the light shining on the solar cell.

9 **Record Data** Record your observations of what happened when the light was blocked.

Communicate Information

1. Did your results match your prediction? Explain.

2. **Construct an Explanation** Based on your observations, what do you think the purpose of a solar cell is?

Name ______________________ Date __________ EXPLAIN

Obtain and Communicate Information

Vocabulary

Use these words when explaining how light transfers energy.

photon | electromagnetic spectrum | visible spectrum
solar cell

Light Energy

Watch *Light Energy* on the amazing properties of light. Answer the question after you have finished watching.

1. Light energy can be transformed into what form of energy by a solar power plant?

____

____

Light and the Electromagnetic Spectrum

Read pages 336–339 in the ***Science Handbook***. Answer the questions after you have finished reading.

2. Describe light's electric and magnetic fields.

____

____

____

____

____

3. Light is both a wave and a particle. Explain what these particles are.

____

____

____

4. What is the electromagnetic spectrum made up of?

5. What is the name of the band in the electromagnetic spectrum that we can see? What band is just before this band? Just after?

6. What happens to the energy that light carries as you move from radio waves to gamma rays?

FOLDABLES®

Cut out the Notebook Foldables given to you by your teacher. Glue the anchor tabs as shown below. Use what you have learned to explain how light is both a wave and a particle.

Glue anchor tab here

Glue anchor tab here

Energy Transfer Through Matter

Investigate how light transfers energy by revisiting the light portion of this simulation. Answer the question after you have finished.

7. Explain how the light from the laser causes the temperature of particles in the table to increase.

______
______
______
______
______
______

Solar Power

Read *Solar Power* on how light energy can be transformed to electrical energy. Answer the following questions after you have finished reading.

8. What are solar cells?

______
______
______

9. What is the difference between solar cells and solar panels?

______
______
______
______

10. Explain how solar cells use light to transfer energy.

____

____

____

____

____

Science and Engineering Practices

Think about the solar circuit that you built earlier. Tell how you planned and carried out an investigation about light energy transfer by completing the "I can . . ." statement below.

I can ____

____

____

____

 Name ____________ Date ______

Research, Investigate, and Communicate

How Much Energy is Used?

You know that we get a lot of light energy from the Sun, but how much of that energy can we use?

Only about 10 percent of the energy from the Sun that is absorbed at Earth's surface, is transformed into chemical energy by plants in photosynthesis. The rest of the energy is transformed into thermal energy. Answer the following questions.

1. What percentage of energy is transformed into thermal energy?

To calculate percentages, write the percentage as a decimal. (For example, 10 percent (%) is equal to 0.10.) Then, multiply the decimal by the total number, like this: $0.10 \times 68{,}000 = 6{,}800$. So, 10 percent of 68,000 is 6,800. To check your answer, multiply your answer by 10, like this: $6{,}800 \times 10 = 68{,}000$.

2. Suppose that 95,000 energy units come from the Sun. Calculate how many units would be transformed into chemical energy by plants.

---

3. From the last example, how many units are transformed into thermal energy?

---

4. Suppose that 6,400 energy units were transformed to chemical energy by plants. How many energy units originally came from the Sun?

---

5. From the last example, how many energy units were transformed into thermal energy?

Using Lasers

Read *Using Lasers* on how lasers are used in different ways today. Answer the following questions after you have finished reading.

Write About It

Find out more about one of the uses of lasers. You will write an expository essay about one use of lasers. Support your main idea with facts and details. Write a concluding sentence at the end.

Getting Ideas

Brainstorm a list of uses for lasers. Choose one to write about and then do some research. Use the table below to record information that you find.

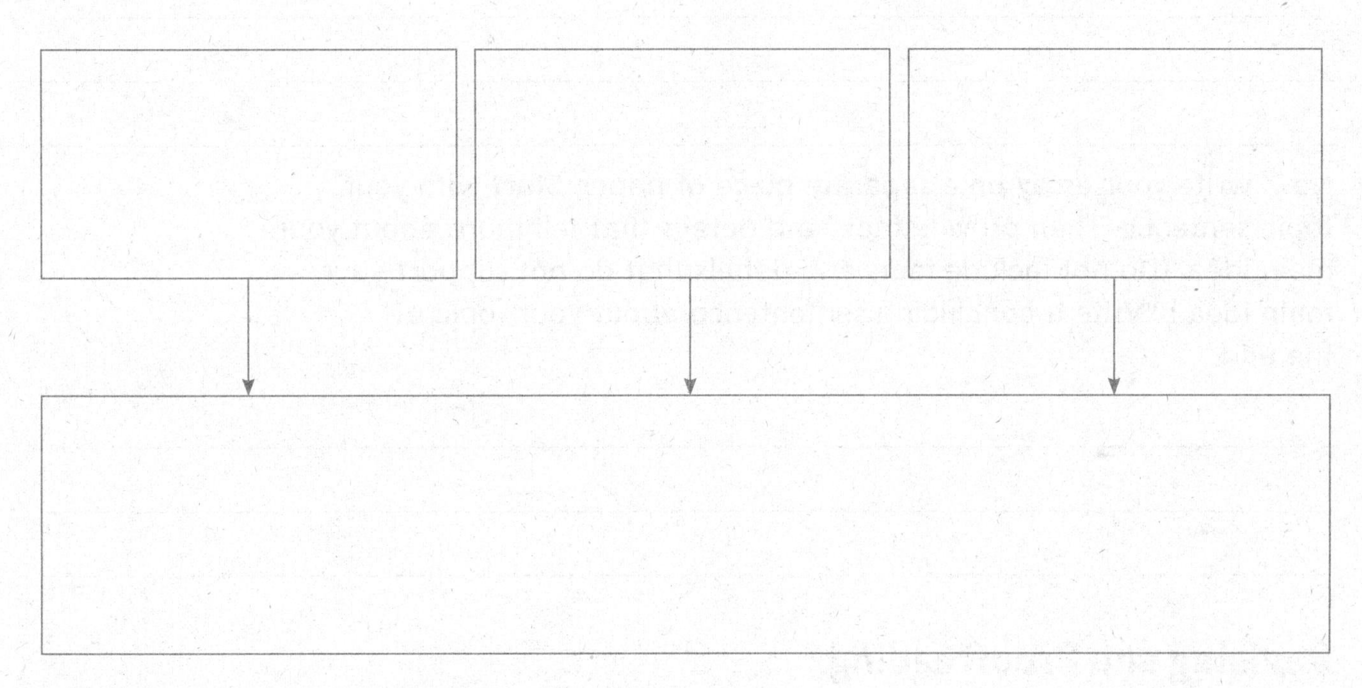

Planning and Organizing

Nick wanted to write about the benefits of laser surgery. Here are the four sentences he wrote. Write *Yes* if the sentence belongs in his essay. Write *No* if it does not.

6. The laser seals the blood vessels when it cuts.

______

7. When lasers are used, there is less blood lost during surgery.

______

8. Scientists used lasers to measure the distance between Earth and the Moon.

______

9. Lasers cut down on the risk of getting an infection from surgery.

______

Drafting

Write a sentence to begin your essay. State your topic, which is the use of lasers you chose to write about. Tell your main idea about this topic. This sentence is your topic sentence.

______

______

______

Now, write your essay on a separate piece of paper. Start with your topic sentence. Then provide facts and details that tell more about your main idea. (Do not include facts and details that do not support your main idea.) Write a concluding sententence about your topic at the end.

______

______

______

Revising and Proofreading

Now revise and proofread your writing. Ask yourself:

- Did I clearly state my main idea about a use of lasers?
- Did I support my main idea with facts and details?
- Did I reach a conclusion at the end?
- Did I correct all mistakes?

Performance Task
A Bright Idea!

As a photonics engineer, you will design an experiment to prove that light transfers energy.

Materials

- ☐ safety goggles
- ☐ 2 identical cups of cold water
- ☐ 2 sheets of black construction paper
- ☐ 2 thermometers
- ☐ desk lamp
- ☐ stopwatch

Write a Hypothesis Write a hypothesis about how light energy from a lightbulb will affect the temperature of water.

____

____

____

____

Carry Out an Investigation

BE CAREFUL Wear safety goggles. Do not spill water near lamp.

1. Design your investigation using the materials listed. List the steps that you will follow to complete your investigation.

____

____

____

____

____

____

____

____

____

2. Follow the steps above and carry out the investigation.

3 Record Data Make a table to record data as you collect it at regular spans of time.

4 Analyze Data Use a separate sheet of graph paper to make a bar graph to show the different temperatures that were measured during the investigation. Make sure you label the axes and title the graph. Attach your graph paper where indicated on this page. What does your graph show?

Glue your graph here.

Crosscutting Concepts

Energy and Matter

1. How did you measure or observe energy transfer in this activity?

2. What patterns did you see as you measured temperature?

Essential Question

How does light transfer energy?

Think about the photo of a snow-covered landscape at the beginning of the lesson. Using what you have learned in this lesson, explain how this landscape could be affected by light energy transfer.

Science and Engineering Practices

Review the "I can . . ." statement you wrote earlier in the lesson. Explain what you have accomplished in this lesson by completing the "I did . . ." statement.

I did ______________________

 Name ______________________ Date __________

Design Energy Solutions

Energy Transfer Engineering Problem

A group of students are working on an engineering project. They have to design a container that would slow down the transfer of heat. They each have different ideas about how to start the engineering process. This is what they said:

Georgie: *I think we need to start by identifying all the materials and resources we can use.*

Hal: *I think we need to start by defining the engineering problem.*

Martha: *I think we need to start by building a model.*

LaBron: *I think we need to start by brainstorming possible solutions.*

Who do you agree with the most? ______________________

Explain why you agree.

____

____

____

____

____

____

____

Name ______________________ Date ____________ ENGAGE

Science in Our World

Look at the photo of the scientist and tents in a harsh environment. What questions do you have?

____

____

____

Read about a thermal engineer and answer the questions on the next page.

STEM Career Connection

Thermal Engineer

As a thermal engineer, I have studied this building and labeled the different temperatures I detected. These observations can show where heat is being lost from this building. Using the data collected, we can make adjustments to the heating system and insulation to reduce the amount of thermal energy transfer that occurs. This will help reduce the energy needed to keep this building warm in the winter and cool in the summer.

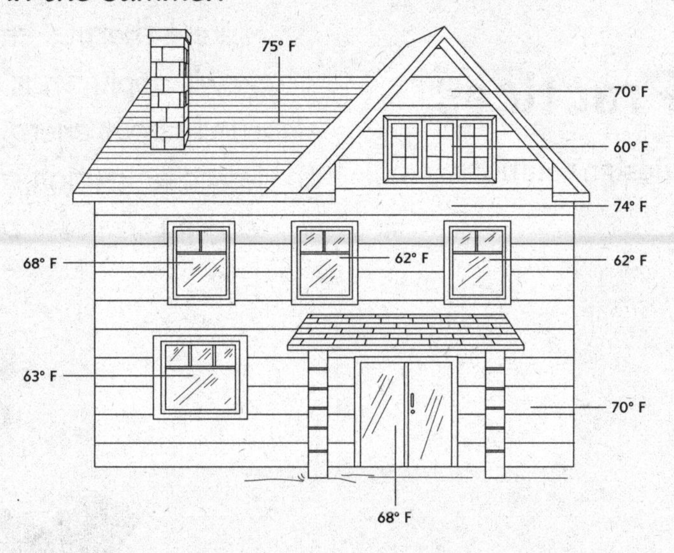

Thermal engineers use their knowledge of thermal energy transfer and the design process to make useful products.

MALIK
Photonics Engineer

 Name ______________________ Date __________

1. On the picture of the house, circle the areas that are showing the most heat loss.

____

____

____

2. What might you do to reduce the amount of heat lost from one of the areas you listed above?

____

____

____

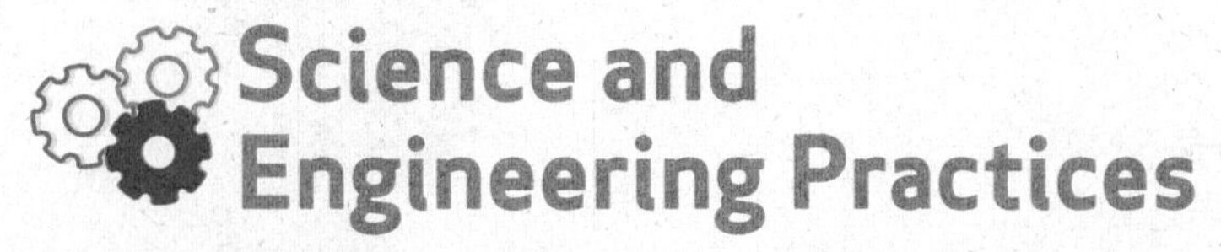

Essential Question

What kinds of problems can be solved by understanding energy transfer?

____

____

____

Science and Engineering Practices

I will construct explanations and design solutions.

Like a thermal engineer, you will apply what you have learned about energy transfers to design a solution to a thermal energy transfer problem.

Inquiry Activity
It's Too Loud in Here!

Materials

- [ ] 4 paper towel rolls
- [ ] tissue paper
- [ ] plastic bubble wrap
- [ ] thick cloth (wool or cotton)
- [ ] wind-up clock

Can you make a decision about which material does a better job of reducing sound transfer? You will explore how different materials affect the ability of sound to move from place to place.

Make a Prediction Which material do you think will best reduce sound energy transfer better: tissue paper, plastic bubble wrap, or thick cloth?

____

____

____

Carry Out an Investigation

1. Observe each of the materials, and then set the four paper towel rolls in front of you.
2. One paper towel roll will be left empty. Set that roll off to the side.
3. Stuff tissue paper into one paper towel roll, filling it completely.
4. Fill the two remaining paper towel rolls. Put plastic bubble wrap in one and thick cloth in the other.
5. **Test Your Solution** Test your devices by placing the wind-up clock at one end of the paper towel roll while you listen at the other end. Start with the empty paper towel roll.
6. **Record Data** Record your observations about how well each material reduced the transfer of sound energy. Use the table on the next page.

Name ______________________ Date ____________

Tube	Observation			
empty				
tissue paper				
plastic bubble wrap				
thick cloth				

Communicate Information

1. Which material was best at reducing sound energy transfer?

____

____

2. How did your prediction compare with the results of your investigation?

____

____

____

3. How do you think this investigation relates to how engineers solve problems?

____

____

____

Name ______________________ Date ____________

Obtain and Communicate Information

Vocabulary

Use these words when explaining how to design energy solutions.

design process	constraint	prototype
	criteria	

Keep It Cool

Read *Keep It Cool* on how engineers have solved a problem involving keeping things cold. Answer the following questions after you have finished reading.

1. What type of energy transfer causes the ice in your drink to melt?

____

____

____

2. Explain how you can tell which way energy is flowing as the ice in your drink melts.

____

____

____

____

3. Draw and label a diagram to explain how an insulator works.

EXPLAIN Name ____________________ Date __________

The Design Process

Explore the Digital Interactive *The Design Process* on the steps engineers use to find solutions to problems. Answer the following questions after you have finished.

4. Fill in the table below by explaining what happens at each step of the design process.

Step	Explanation
1. Identify the Problem	
2. Develop Solutions	
3. Choose a Solution	
4. Build a Prototype	

Name ______________________ Date __________ EXPLAIN

5. Test the Prototype	
6. If Test Is Not Successful	
7. Make a Final Design and Communicate the Results	

5. Why are prototypes an important part of the design process?

____

____

____

____

EXPLAIN Name ______________________ Date __________

The Design Process in Action

Read *The Design Process in Action* on how engineers designed a solution to a problem. Answer the following questions after you have finished reading.

6. Match each step of the design process with the steps taken to design a new airplane.

1. Identify a Problem	• **Replace the engine with a bicyclist.**
2. Develop Solutions	• **Use a new engine. Replace the engine with a bicyclist. Use a solar panel.**
3. Choose a Solution	• **The *Gossamer Penguin***
4. Build a Prototype	• **Find an alternate form of power.**
5. Test the Prototype	• **The *Gossamer Albatross***
6. Make a Final Design	• **Fly the Gossamer Albatross across the English Channel.**

7. What is the design process?

____

____

8. Does the design process ever end? Explain your answer.

____

____

____

____

Design Solutions

Use different sets of criteria and constraints to choose the best solution to design problems.

Design Problem A vacuum company is getting customer feedback that their original vacuum needs to be improved. Engineers have brainstormed three possible solutions to this problem listed below.

Solution #1	Solution #2	Solution #3
Replace the materials in the original vacuum with lighter-weight plastic.	Make a vacuum that is lightweight, battery-operated, has breakaway parts, and includes a telescoping handle.	Make a computerized robotic vacuum that climbs stairs and changes tasks by voice command.

The manufacturer has discovered that they have three different user groups. Each group has different sets of criteria for and constraints to the design. Your job is to use each set of criteria and constraints for each of the user groups to choose the best solution to the design problem.

9. User Group #1

Criteria: Vacuum must be lightweight (less than 20 pounds) and easy for the average adult to use.

Constraints: The design of the original vacuum must not change. Users like the current design.

Which solution best fits these criteria and constraints? Explain your decision.

____

____

____

____

10. User Group #2

Criteria: Vacuum must be usable by people who have physical limitations and cannot lift more than 5 pounds.

Constraints: Vacuum must not contain multiple parts that can be lost.

Which solution best fits these criteria and constraints? Explain your decision.

______

______

______

______

11. User Group #3

Criteria: Vacuum must be easy to use on stairs and furniture.

Constraints: Vacuum must be able to work where it cannot be plugged in with a cord.

Which solution best fits these criteria and constraints? Explain your decision.

______

______

______

______

______

12. What role do criteria and constraints play in the design process?

____

____

____

____

Science and Engineering Practices

Think about what you learned about designing energy solutions when you discovered which material quieted sound the best. Tell how you can apply scientific ideas to solve design problems by completing the "I can . . ." statement below.

I can ____

____

____

____

 Name ______________________ Date __________

Research, Investigate, and Communicate

Technology and Energy Transfer

Watch *Technology and Energy Transfer* on how engineers developed different types of devices that use energy transfer. Answer the question after you have finished watching.

1. Imagine how your life would be different without some of the technology we have that was developed by scientists and engineers. Describe how you would communicate differently with your friends and family if cells phones had not been invented.

____

____

____

____

____

____

Changing Technology Over Time

Choose an everyday technology that has been continually improved over time. You will use this technology as your research topic. Remember that "technology" does not have to be high-tech. Your topic could be something as simple as pencils or clothespins.

Use different resources to research how the design process has helped the technology change over time. You will present your findings to the class in an illustrated presentation.

2. Take notes as you research and prepare for your presentation.

____

____

____

____

____

FOLDABLES®

Cut out the Notebook Foldables tabs given to you by your teacher. Glue the anchor tabs as shown below. Use what you have learned to describe a problem involving energy transfer that you might expect to have in the picture. How might you solve that problem?

Performance Task
It's Too Cold in Here!

Revisit the photo of the scientist walking toward the tents in the snow. Look at the photo of the thermal engineer. You will research materials that could be used to reduce thermal energy transfer. You will then develop a model of a device that can help people to survive in the wilderness. Your device will need to solve the problem of reducing the amount of heat lost in a tent, but the device also cannot be heavy to carry.

Make a Model

1. Think about the type of materials that would be helpful in preventing heat loss. List each material you would use and its purpose.

2. Draw a diagram of your solution. Label your diagram showing the materials you will use.

Crosscutting Concepts
Energy and Matter

3. How does your design reduce the amount of thermal energy transfer from the tent?

______

4. Explain how you would use your model to build a real prototype.

______

5. How could you test your design?

______

6. What was the constraint in designing your solution?

7. How would your design change if you had no constraints?

8. For your solution to be successful it had to meet a criterion. What was the criterion for this solution?

9. How would you use the engineering design process to help you improve the tent?

 Name ______________________ Date __________

Essential Question

What kinds of problems can be solved by understanding energy transfer?

Think about the photo of the scientist with tents in the snow at the beginning of the lesson. Explain how the thermal engineer's understanding of energy transfer helped solve a problem.

____

____

____

____

____

____

Science and Engineering Practices

Review the "I can . . ." statement you wrote earlier in the lesson. Explain what you have accomplished in this lesson by completing your "I did . . ." statement.

I did ______________________________________

____

____

____

____

____

Transfer of Energy

Performance Project
Design a Windmill

Construct a model to demonstrate how energy can be transferred from place to place. Wind energy engineers use what they know about energy transfer to design wind turbines that generate as much electricity as possible. They collect data to improve wind turbine design.

Now it is your turn to be an engineer. You will build and test a windmill. Then you will use your knowledge of energy transfer to improve your design.

Design a model windmill to lift a paper clip attached to it as it spins. Your goal is to determine what type of windmill design will lift the paper clip the fastest.

Define a Problem What variable do you think you can change to lift the paper clip faster?

______

______

______

Materials

- ☐ safety goggles
- ☐ pencil
- ☐ milkshake straw
- ☐ scissors
- ☐ masking tape
- ☐ available classroom materials
- ☐ 30-cm piece of thread
- ☐ paper clip
- ☐ stopwatch

Make a Model

BE CAREFUL Wear safety goggles during this activity.

1. Using the Internet or other available research materials, research designs of wind turbines. Use the space below to take notes.

______

______

______

 Name ______________________ Date __________

Transfer of Energy

2 Use what you found in your research to draw a design for a windmill in the box below.

Design 1	Design 2

3 Use scissors to cut the milkshake straw to about 2 cm shorter than the pencil. Put the pencil through the straw to form the bearing on which the windmill will rotate.

4 Use available classroom materials to create windmill blades based on your design. Use masking tape or other classroom materials to attach the blades to the milkshake straw near one end. Make sure the blades are equally spaced around the straw.

5 Tie the paper clip to the thread. Tape the other end of the thread to the straw at the end opposite the blades.

6 **Test** Hold the ends of the pencil and blow on the windmill blades. Experiment with where and how to blow on it to get it to spin. Then, have a partner use the stopwatch to time how long it takes to lift the paper clip to the pencil. Record your results in the table below. Repeat the test two more times and calculate the average.

7 How can you change the variable that you identified to lift the paper clip faster? Draw your new design above in the box labeled Design 2. Then, construct and test your design by repeating step 6.

	Time to Lift Paper Clip (s)			
	Trial 1	**Trial 2**	**Trial 3**	**Average**
Design 1				
Design 2				

8 How could you use a windmill to generate electricity? Draw a labeled diagram to show how your design could be used to generate electricity.

Communicate Information

1. How was the energy from your breath used to raise the paper clip?

2. **Construct an Explanation** Which design worked better? How do you know?

Explore More in Our World

Did you learn the answers to all of your questions from the beginning of the module? If not, how could you design an experiment or conduct research to help answer them?

 Name ________________ Date ________

Structures and Functions of Living Things

Science in Our World

Look at the eye of the cuttlefish. Why do you think the cuttlefish has pupils that are shaped like a "W"? What questions do you have about the picture?

____

____

____

____

____

Key Vocabulary

Look and listen for these words as you learn about the structures and functions of living things.

adaptation	image	nervous system
peripheral nerve	reflection	refraction
respiration	response	stimulus
structural adaptation	tropism	

Name ______________________ Date __________

How are the shapes and sizes of animal eyes related to their functions?

HIRO
Ocean Engineer

STEM Career Connection

Marine Biologist Field Notes

Date: June 8, 2012 | Time: 1:12 pm

Species Observed: Lanternfish | Number Observed: 85 individuals

Depth: 1,100 meters | Average Size: 12 centimeters

Notes about Structures: Lanternfish have relatively large eyes and use bioluminescence to glow in the dark.

Draw and label a diagram to show how you think the shape of an animal's eye helps it see.

Science and Engineering Practices

I will engage in argument from evidence.

I will develop and use models.

 Name ______________________ Date ____________

Structures and Functions of Plants

Plant Parts

Plants are made up of different parts that help the plant live in its environment. Put an X in all of the boxes that contain a part that could be found on a plant.

Roots	Leaves	Bark
Flower	Nuts	Seeds
Spines	Stems	Trunk
Branch	Pine needles	Tubes that carry water
Root hairs	Fruit	Waxy coating

Explain your thinking. How did you decide which things were parts of a plant?

____

____

____

____

____

Science in Our World

Look at the photo of the plant. What questions do you have?

____

____

____

____

Read about a botanist and answer the questions on the next page.

STEM Career Connection

Botanist

Today I presented the results of my most recent research to the other scientists at the plant biology conference. It was very rewarding to finally share my work and see the reactions and questions that the other scientists had. My work has answered some questions, but it has led to others that we will begin researching soon.

My research studied the effects of air pollution on crops. It required me to be in the field for several hours each week, studying the crops while they were growing. Over the winter, I analyzed the data that I had collected and wrote reports and scientific journal articles. The study took several years to complete. I worked together with other scientists at the university and at the state department of agriculture.

This is the perfect lesson for a botanist because it's all about plants!

OWEN
Entomologist

 Name ______________________ Date __________

1. Using clues from the botanist's journal entry, what do you think a botanist does? What clues did you use?

2. When did the botanist analyze collected data? Why do you think the data was analyzed at this time?

Essential Question

How do plant structures help them survive and reproduce?

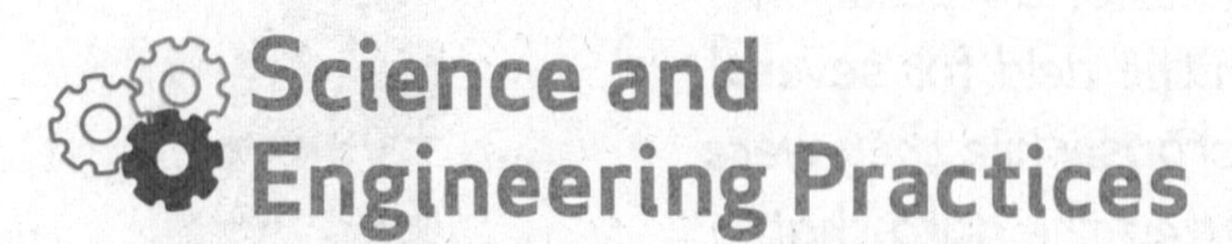

Science and Engineering Practices

I will engage in argument from evidence.

Like a botanist, you will make an argument about plant structures and functions using evidence, data, and models throughout this lesson.

Name ______________________ Date __________ EXPLORE

Inquiry Activity

Movement of Water in Plants

How does water move through a plant?

Make a Prediction You will observe four plants of the same type. You will give each plant the same amount of water and then enclose each in a plastic bag from the stem down. Make a prediction about how water will move through the plants.

____

____

Materials

- ☐ water
- ☐ graduated cylinder
- ☐ 4 plants in containers
- ☐ 4 plastic bags
- ☐ string
- ☐ masking tape
- ☐ marker
- ☐ pan balance
- ☐ light source

Carry Out an Investigation

1. Use the graduated cylinder to measure and pour 60 milliliters of water into each plant container.
2. Place each plant into a bag. Make sure each plant is upright.
3. Use a piece of string to carefully tie the bag closed around the stem of each plant. Most of the stem and leaves of the plant should be open to the air.
4. Place a piece of masking tape on each bag. With a marker, label each plant consecutively, 1-4.
5. **Record Data** Use the pan balance to determine the mass of each plant in grams. Record the mass of each plant in the table.

	Plant #1 (under light)	Plant #2 (under light)	Plant #3 (out of light)	Plant #4 (out of light)
Plant's initial mass:				
Plant's mass: 24 hours				
Plant's mass: 48 hours				
Plant's mass: 72 hours				

6 Place Plant #1 and #2 under the light source. Place Plant #3 and #4 in positions that do not get light.

7 **Record Data** After 24 hours, use the pan balance to find and record the mass of each plant. Return each plant to its last position.

8 **Record Data** Repeat step 7 after 48 hours and 72 hours.

9 **Analyze Data** Use a sheet of graph paper to make a line graph showing changes in each plant's mass. Label your graph's X and Y axes and title your graph. Paste your graph where shown.

Communicate Information

1. In your investigation, what variable did you change?

____

____

2. How did your plants' masses changed over time?

____

____

3. Why do you think the plants' mass changed?

____

____

____

4. How can this activity help show that water moves through a plant?

____

____

____

____

____

____

Glue your graph here.

FOLDABLES®

Cut out the Notebook Foldables tabs given to you by your teacher. Glue the anchor tabs as shown below. Use what you have learned to describe external and internal structures of plants and their functions. Use the pictures to help you.

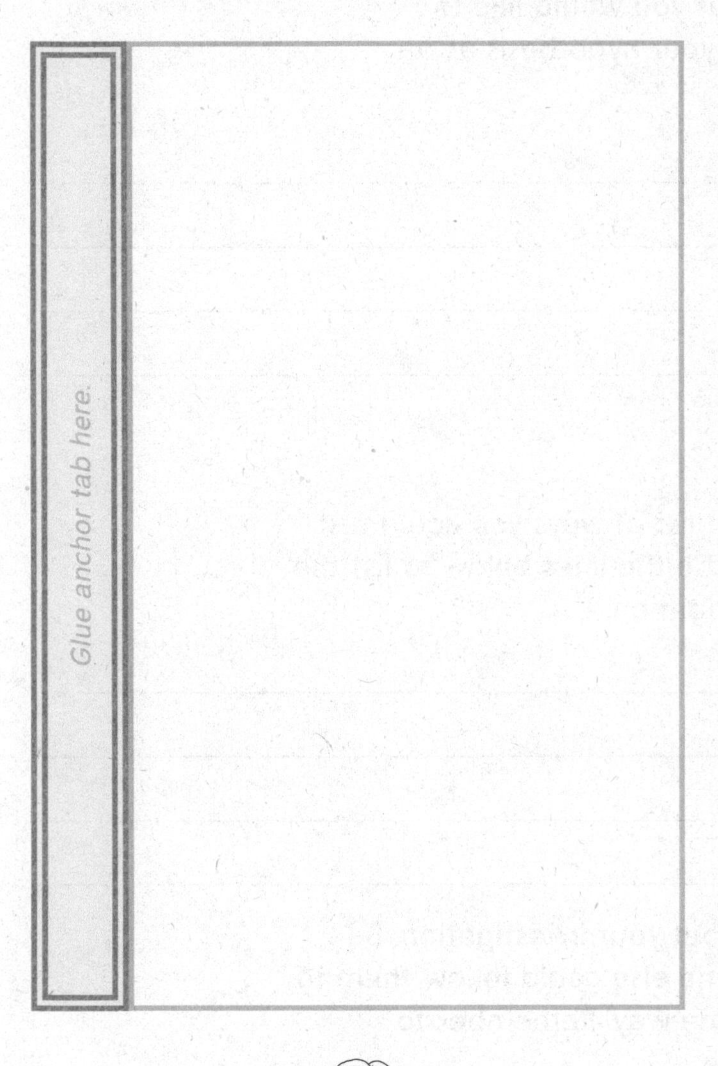

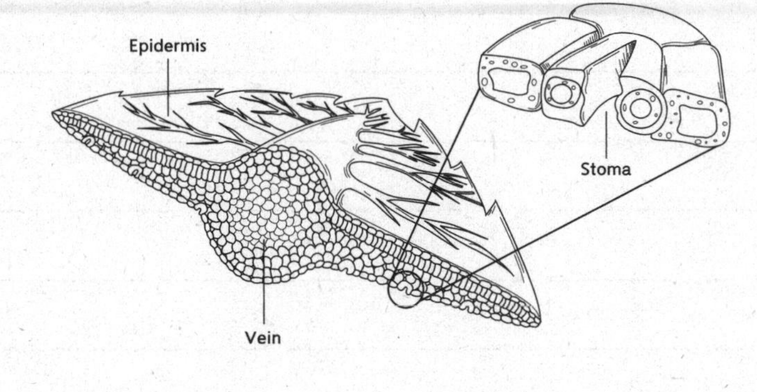

ELABORATE Name ______________________ Date __________

Inquiry Activity
Design an Experiment

In the following activity, you will design an experiment to test how a plant would react to a water source.

Write a Hypothesis Develop a hypothesis that you would like to answer by completing an experiment. Write your hypothesis as an "If..., then..." statement.

____

____

____

____

Carry Out an Investigation

1. On a separate sheet of paper, brainstorm a list of ways you could test your hypothesis. Select one of your ideas. Use the lines below to list the materials that you will need for your investigation.

____

____

____

____

2. Write the steps you would follow to carry out your investigation. Be specific with your directions so that someone else could follow them to complete the investigation in the exact same way. Remember to inlcude all of the materials you listed above.

____

____

____

____

____

____

Name ____________________ Date __________ ELABORATE

3. How does your experiment allow you to construct an argument with evidence?

____

____

____

____

We know that plants respond to stimuli in their environment. Did you know that we can use math to help us understand some of these responses? Plants respond to sunlight by growing towards it. This causes the plant's stem to form an angle with the ground. An angle is formed by two lines that meet at a common endpoint. The stem meets the ground at this endpoint.

There are three main types of angles. A right angle has a square corner where the lines meet. An obtuse angle has a wider opening than a right angle. An acute angle has a smaller opening than a right angle.

4. Circle the diagram that shows an acute angle.

5. Color the sky of the diagram that shows an obtuse angle.

6. Do either of the plants have right angles? How do you know?

____

____

____

Performance Task

How Do Plants Respond to Changes in Their Environments?

Materials

- ☐ safety goggles
- ☐ shoe box with lid
- ☐ scissors
- ☐ ruler
- ☐ cardboard
- ☐ tape
- ☐ potted plant
- ☐ camera

As a botanist, you will gather data that you can use to make an argument about how a plant responds to changes in its environment.

Write a Hypothesis Can a plant grow around obstacles in order to meet its needs? In this activity you will build an obstacle course that a plant must overcome in order to reach light.

Use what you have learned to develop a hypothesis to explain how plant structures change to help the plant survive when light is blocked. Remember to write your hypothesis as an "If..., then..." statement.

____

____

____

____

Carry Out an Investigation

BE CAREFUL Wear safety goggles and handle the scissors with care.

1. Use the drawing below as a model.
2. On the right-hand side of one end of the shoe box, carefully cut an opening. The opening should be approximately 5 centimeters (cm) wide by 5 cm tall.
3. Cut two dividers from the cardboard. Make them as tall as the shoe box, but about 5 cm less wide.
4. Place the dividers upright inside the box. Tape the first divider to the same side as the opening you cut, but closer to the opposite end of the box. Tape the other divider a few inches away from the first divider, but on the opposite side of the box.

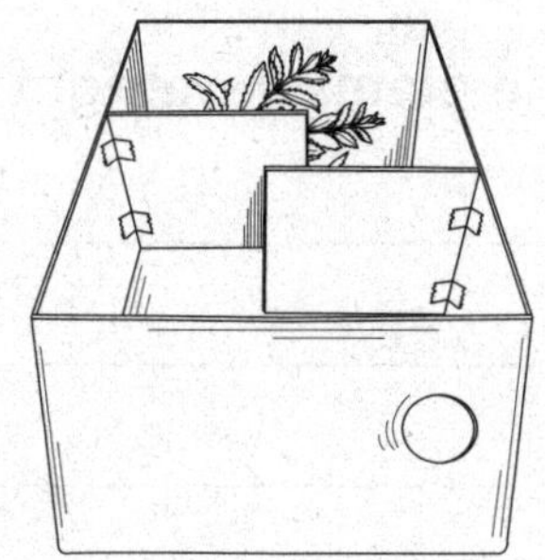

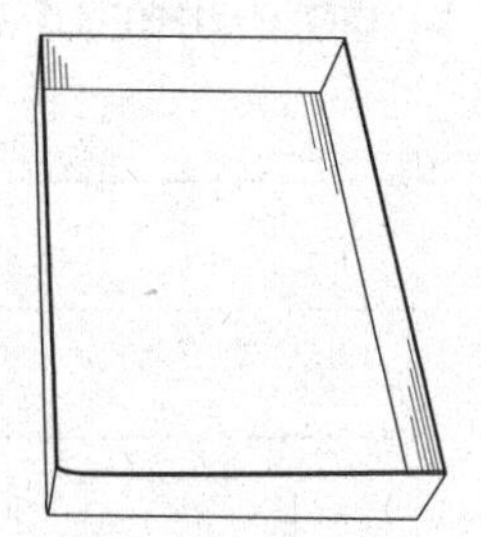

Name ______________________ Date __________ EVALUATE

5 Place the plant in the end of the box opposite the opening.

6 **Record Data** Use the ruler to measure the height and width of the plant. Record your data in the table. Include any observations you make about your plant.

Day	Plant Height (cm)	Plant Width (cm)	Observations
0			
3			
6			
9			

7 Sketch your box and plant. Include measurements in centimeters for the box and the plant in your drawing.

8 Put the lid on the box and turn the opening toward the window or a bright light source.

9 Remove the lid and take photos of the plant. Observe and measure the growth of your plant. Record your measurements and observations in the data table.

10 After recording your measurements, give your plant some water and replace the lid. Return your shoe box to its original position if you moved it.

11 Repeat steps 9 and 10 every three days for several weeks.

EVALUATE Name ______________________ Date __________

Communicate Information

1. Look back at your photos, measurements, and observations. Describe how your plant has changed over time.

____

____

____

2. **Make an Argument** How did the plant's structures change to help the plant survive? Use the data you collected as evidence to support your argument.

____

____

____

____

____

____

____

____

____

3. How did this investigation support your hypothesis?

____

____

____

____

____

____

Name ______________________ Date ____________ EVALUATE

Essential Question

How do plant structures help them survive and reproduce?

Think about the photo of the bending plant you saw at the beginning of the lesson. Using what you have learned, explain how the bending plant uses its structures to help it survive.

____

____

____

____

____

Science and Engineering Practices

Review the "I can . . ." statement you wrote earlier in the lesson. Explain what you have accomplished in this lesson by completing your "I did . . ." statement.

I did ____

____

____

____

____

 Name ______________________ Date __________

Structures and Functions of Animals

Animal Parts

Animals are made up of different parts that help the animal live in its environment. Put an X in all of the boxes that contain a part that could be found on an animal.

Ear	Shell	Claw
Heart	Leaf	Feather
Tentacle	Tail	Fur
Roots	Lungs	Skin
Teeth	Antennae	Wings
Fin	Beak	Seeds

Explain your thinking. How did you decide which things were parts of an animal?

____

____

____

____

____

Name ______________________ Date ____________ ENGAGE

Science in Our World

Look at the photos of different animal body structures. What questions do you have?

____

____

____

____

____

Read about a zoologist, and answer the questions on the next page.

STEM Career Connection

Zoologist

After many years of watching freshwater mussel populations decline, we are finally seeing them begin to increase. Freshwater mussels are mollusks that are similar to clams and oysters. They have two shells that are connected on one side. They live in freshwater rivers, streams, lakes, and ponds. Mussels filter the water and eat algae, bacteria, and other small organisms. Since mussels act as filters, they can help to improve water quality.

Unfortunately, many mussel species are threatened or endangered. Mussels are very sensitive to water pollution. They have also been negatively impacted by non-native species and dams.

Recently, zoologists have begun to cultivate mussels. We help the mussels reproduce and grow in captivity, and then we release them. I am happy to report that we have seen increased mussel populations in some locations!

OWEN
Entomologist

Name ______________________ Date ____________

1. What does an increase in the mussel population indicate about water quality?

____

____

____

____

2. Using text clues in the last paragraph, what do you think the word *cultivate* means?

____

____

____

Essential Question

How do animal structures help them survive?

____

____

____

____

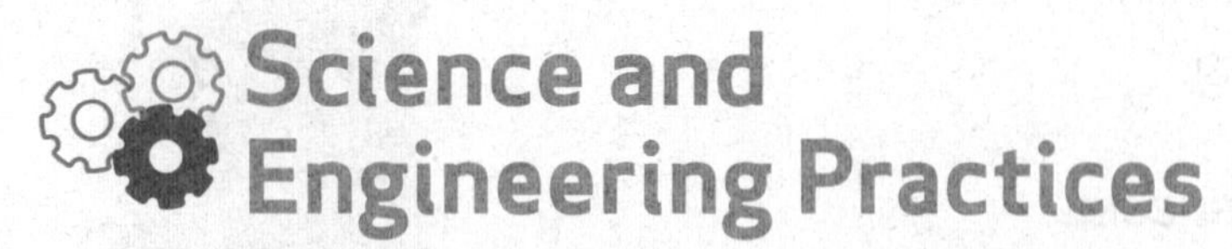

I will engage in argument from evidence.

Name Savannah Date

EXPLORE

Inquiry Activity

Put Your Best Foot Forward

How do different foot structures help animals survive in their environment?

Make a Prediction Look at the tongue depressor, fork, and tweezers. Make some predictions about which one will be best used for swimming. Which tool will be best used for picking things up? Which tool will be best for digging in debris to find food?

I predict that the tongue depressor will be best for swimming. I predict that the tweezers will be best for picking up things. I predict that the fork will be best for digging.

Materials

- ☐ safety goggles
- ☐ 2 plastic containers
- ☐ water
- ☐ pea gravel
- ☐ colored gravel
- ☐ tongue depressor
- ☐ fork
- ☐ tweezers
- ☐ colored pencils

Carry Out an Investigation

BE CAREFUL Wear safety goggles during the activity.

1. Place the two plastic containers on the floor. Fill one container half full with water.
2. Fill the second plastic container one-fourth of the way full with pea gravel.
3. Place a handful of the colored gravel into each container and spread the gravel around.
4. Observe the different tools. They represent different types of feet you might find on birds.
5. **Record Data** Experiment with the tools to determine which would best help a bird to grasp and pick up food (colored gravel), swim in water, and dig through debris (pea gravel). In the table on the next page, rate each tool from 1 to 3 based on how well it performed in each test. 1 is good, 2 is ok, and 3 is poor.

 Name Savannah Date

Tools	Type of Activity		
	Digging for food in debris	Paddling through water	Picking up and grasping food
Tongue depressor	3	2	2
Fork	3	1	3
Tweezers	1	2	3

6 **Analyze Data** In the data table, shade each box in the Tools column as follows: Use red to shade the tool that best helped grasp and pick up food. Use blue to shade the tool that would make swimming easier. Use green to shade the tool that made it easiest to dig through debris looking for food.

Communicate Information

1. How did your prediction compare with your results?

My prediction was right because the tongue depressor was great for swimming, the fork was great for digging, and tweezers were great for picking up things.

2. Why do you think it is helpful for similar animals, like birds, to have different types of body structures?

It is helpful for birds to have different types of body structures because some birds live in different enviorments, whitch requiers different body structures.

Structural Adaptations

Read pages 110, 111, and 113 in the *Science Handbook*. Answer the following questions after you have finished reading.

6. Explain what structural adaptations are and how they help animals survive.

7. Explain how camouflage is a structural adaptation.

8. **Look at the image of the hummingbird and the tiger on page 113 of the *Science Handbook*. Explain how these animals' mouths are structurally adapted for the food they eat.**

How Animals Survive

Explore the Digital Interactive *How Animals Survive* on how animals use their structures to survive in their habitats. Answer the questions after you have finished.

9. Identify several internal structures that some organisms have. Explain how those internal structures help the organism to survive.

10. Identify several external structures that some organisms have. Explain how those external structures help the organism to survive.

FOLDABLES®

Cut out the Notebook Foldables tabs given to you by your teacher. Glue the anchor tabs as shown below. Use what you have learned to describe the function of internal and external structures of a deer and a beetle.

Glue anchor tab here.

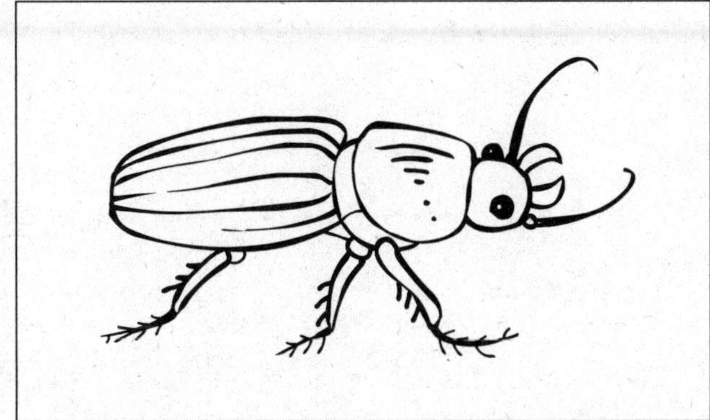

 Name ______________________ Date __________

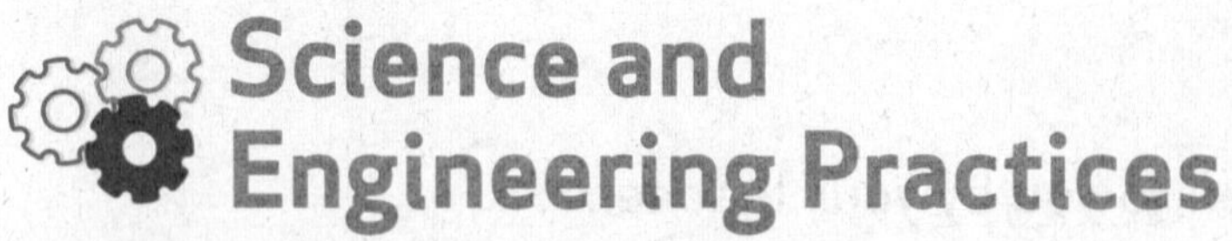

Science and Engineering Practices

Use examples from the lesson to explain what you can do!

Think about the evidence you have collected about how body structures help animals survive. Complete the "I can . . ." statement below to tell how you can construct an argument using evidence.

I can ____

____

____

____

____

____

____

____

Date ______

EVALUATE

erformance Task

ıe Model Is Afoot!

ogist, you will use your knowledge of animals' body ;s to design a model of an animal foot. You will ıe foot to be used for a purpose you choose.

Materials
- ☐ rubber glove
- ☐ available classroom materials

a Model

ıink about the different animals that you have observed ın this lesson. Choose one animal that has feet. Write the name of the animal.

______

2 Explain what you would like this animal's foot to be able to do.

______

______

______

______

______

3 Brainstorm solutions to the problem you identified in step 2.

______

______

______

______

______

4 Choose one solution and explain how you will model it.

______

______

______

______

5 Use a rubber glove to represent your animal's foot. Use available classroom materials to model the structural adjustment you want to make. Draw a diagram of your model and label the structures. Include a description of how the foot functions.

Crosscutting Concepts

Systems and System Models

1. Explain how your animal, with its structural change, would interact with its environment.

2. **Make an Argument** Explain how the change you made to the model foot (the glove) achieved its purpose. Then make an argument for how this adjustment would help the animal survive in its environment. Use evidence you have collected throughout this lesson to support your argument.

Name ______________________ Date ____________

Essential Question

How do animal structures help them survive?

Think about the photos of different animals at the beginning of the lesson. Explain how those animals use their structures to survive in their environment.

____

____

____

____

____

____

____

Science and Engineering Practices

Now that you're done with the lesson, share what you did!

Review the "I can . . ." statement you wrote earlier in the lesson. Explain what you have accomplished in this lesson by completing the "I did . . ." statement.

I did ______________________________

____

____

____

____

____

____

 Name ______________________ Date __________

Information Processing in Animals

Animal Senses

Three friends were talking about how animals use the information they get from their senses. They each had a different idea. This is what they said:

Melinda: *I think each sense organ processes the information from our senses. For example, the nose tells you what the smell is.*

Ralph: *I think the brain processes all the information from our senses. For example, the brain tells you what it is you are seeing.*

Nate: *I think the nerves in our sense organs process the information from our senses. For example, nerves in our fingers tell us what we are touching.*

Who do you agree with the most? ______________________

Explain why you agree.

____

____

____

Name ______________________ Date __________

Science in Our World

Look at the photo of the large cat hunting at night. What questions do you have?

____

____

____

____

____

____

Read about a wildlife biologist, and answer the questions on the next page.

STEM Career Connection

Wildlife Biologist

Today was a rewarding day! I got to see a brown bear that I had been studying and observing for several months released back into the wild. It is my job to study the behavior of animals that have been taken into captivity due to injury or illness. After the animals have recovered, I make sure that their behavior is normal for a wild animal of their species. One important part of my job is making sure that the animal is able to find food in the wild. This brown bear was used to being fed by his caretakers since a leg injury prevented him from hunting and gathering food on his own. After his leg had healed, it took him several months to learn to hunt again.

Name ____________________ Date __________

1. Why might an animal that has been living in captivity not be able to find food in the wild?

2. What senses might a brown bear use to hunt? Explain.

Essential Question

How do animals sense and respond to information?

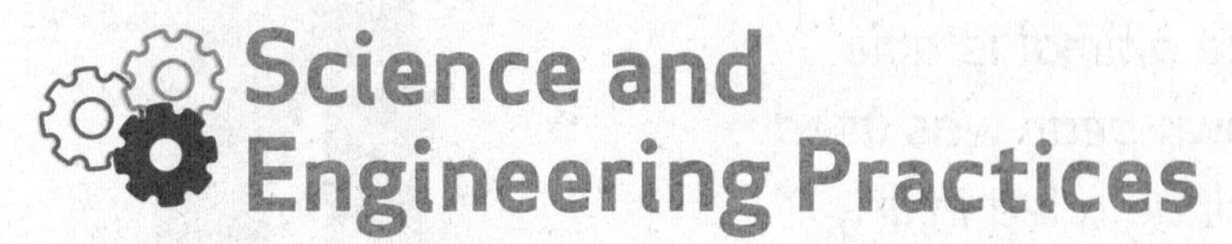

Science and Engineering Practices

I will develop and use models.

Like a wildlife biologist, you will develop and use a model to help you compare animal senses.

Name ______________________ Date __________ EXPLORE

Inquiry Activity
Sense of Touch

How sensitive is your sense of touch? In this activity you will order samples of sandpaper by roughness while blindfolded.

Materials
- ☐ 3 sandpaper samples of different grades
- ☐ material for blindfold
- ☐ hand lens

Make a Prediction How can you order sandpaper samples by roughness without using your sense of sight?

____

____

____

Carry Out an Investigation

BE CAREFUL The blindfold should be snug, but not uncomfortable. When the blindfold is in place, do not move around the classroom.

1. Have your partner help you put the blindfold on.
2. Your partner will arrange the sandpaper samples randomly on a desk in front of you, making sure the textured side of the sandpaper is facing up.
3. Use your sense of touch to order the samples of sandpaper from fine to rough.

Sandpaper Order	Description of Texture

Name ______________________ Date __________

4 **Record Data** Remove the blindfold. Turn the sandpaper samples over, keeping them in the same order. You will see a number. Record this number and a description of the texture for each sandpaper sample in the table on the previous page. Use the hand lens to observe the size of the grains on each sheet of sand paper.

5 Switch roles with your partner and repeat steps 1–4.

Communicate Information

1. **Analyze Data** Was your prediction supported by the data you collected? Explain.

2. Compare how the finest sandpaper felt and how it looked.

3. How did your partner's results compare to yours?

4. **Construct an Explanation** How do you think a sense of touch could help an animal survive?

Obtain and Communicate Information

Vocabulary

Use these words when explaining how animals process information.

nervous system	brain	spinal cord
central nervous system	peripheral nerve	sensory organ
	reflex	

Inquiry Activity
Reaction Time

Materials
- ☐ meterstick

You will explore how quickly your nervous system can react to sensory information.

Make a Prediction How fast do you think you can move when you see something?

____

____

____

Carry Out an Investigation

1. Have your partner hold a meterstick at the end with the highest number.
2. Place the thumb and forefinger of your left hand around, but not touching, the end of the meterstick marked with a zero.
3. When your classmate drops the meterstick, try to catch it between your thumb and forefinger as quickly as you can.
4. **Record Data** Use the table on the next page to record where the top of your thumb is when you catch the meterstick. Your measurement should be to the nearest half centimeter. If you miss catching the meterstick for any trial, then record 100 centimeters for that trial.

Name ____________________ Date __________

Trial #	Your Results – Measurement on meterstick (cm)	Partner Results – Measurement on meterstick (cm)
1		
2		
3		
4		

5 Repeat steps 1–4 three more times.

Communicate Information

1. **Analyze Data** Calculate the average measurement for you and your partner. Use a separate sheet of graph paper to make and label a bar graph to show your results. Attach your graph to this page where indicated.

2. Who reacted more quickly? How do you know?

____

____

____

3. Was your prediction correct? Explain.

____

____

____

Glue your graph here.

The Brain and Parts of the Nervous System

Explore the Digital Interactive *The Brain and Parts of the Nervous System* on how sensory information is taken in and interpreted. Answer the questions after you have finished.

4. How do the brain, spinal cord, sensory organs, and peripheral nerves work together to react to sensory information?

5. The brain is often called the control center of the human body. Is this a good comparison? Why or why not?

Structural Adaptations–Animal Senses

Read page 112 in the ***Science Handbook***. Answer the following question after you have finished reading.

6. How are the senses of pit vipers different from yours?

EXPLAIN Name ______________________ Date __________

Animal Senses

Read *Animal Senses* on how animals use their senses to survive.

7. How do animals use their sense of smell?

____

____

____

____

Brain Illumination

Investigate which parts of the brain interpret information from the different sense organs by conducting the simulation.

8. Describe what happens in your nervous system when you smell a bad smell. Use the words *central nervous system* and *peripheral nervous system*.

____

____

____

____

____

____

Science and Engineering Practices

Use examples from the lesson to explain what you can do!

Think about how you developed and used models to learn about animal senses and the nervous system. Tell how you can develop and use models by completing the "I can . . ." statement below.

I can ____

____

____

Research, Investigate, and Communicate

The Nervous System

Read page 391 in the ***Science Handbook***. Answer the following questions after you have finished reading.

1. Use the words *stimulus*, *response*, and *neuron* to describe what happens when you smell a skunk. (Hint: You can use information from the simulation to help you.)

____

____

____

____

____

____

2. What is a reflex? Use the words *stimulus*, *response*, and *neuron* to contrast what you wrote for number 1 with what happens for a reflex. Give an example, and tell how the reflex helps an animal survive.

____

____

____

____

____

____

____

____

____

____

____

Name ____________________ Date __________

FOLDABLES®

Cut out the Notebook Foldables tabs given to you by your teacher. Glue the anchor tabs as shown below. Use what you have learned to explain how the dog processes information using its sensory organs.

Glue anchor tab here

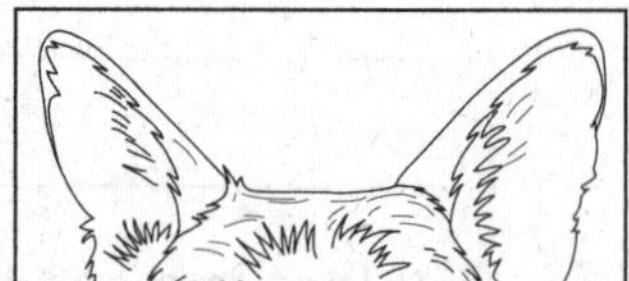

Name ______________________ Date ______________

Performance Task
Comparing Senses

As a wildlife biologist, you will design a model to compare how humans and large cats use their senses to gather information about their environments. You will use the model to demonstrate how large cats sense information differently than humans.

Materials
- ☐ research materials
- ☐ various classroom resources to build a model

Night Vision Goggles

Read *Night Vision Goggles* on a technology that helps humans see at night. Answer the following question after you have finished reading.

Make a Prediction How do you think the senses of a large cat, like the one from the beginning of the lesson, differ from those of a human?

____

____

____

____

Make a Model

1. Think about the large cat from the photo at the beginning of the lesson and the Science File you just read about night vision goggles. Use several different resources to research the senses of large cats. Record your notes on a separate sheet of paper.

2. Use classroom resources to develop a model that shows the differences between human and big cat senses. Sketch and label your model below.

 Name ____________________ Date __________

Communicate Information

1. Analyze Data Compare and contrast human senses with those of the large cat using the Venn diagram.

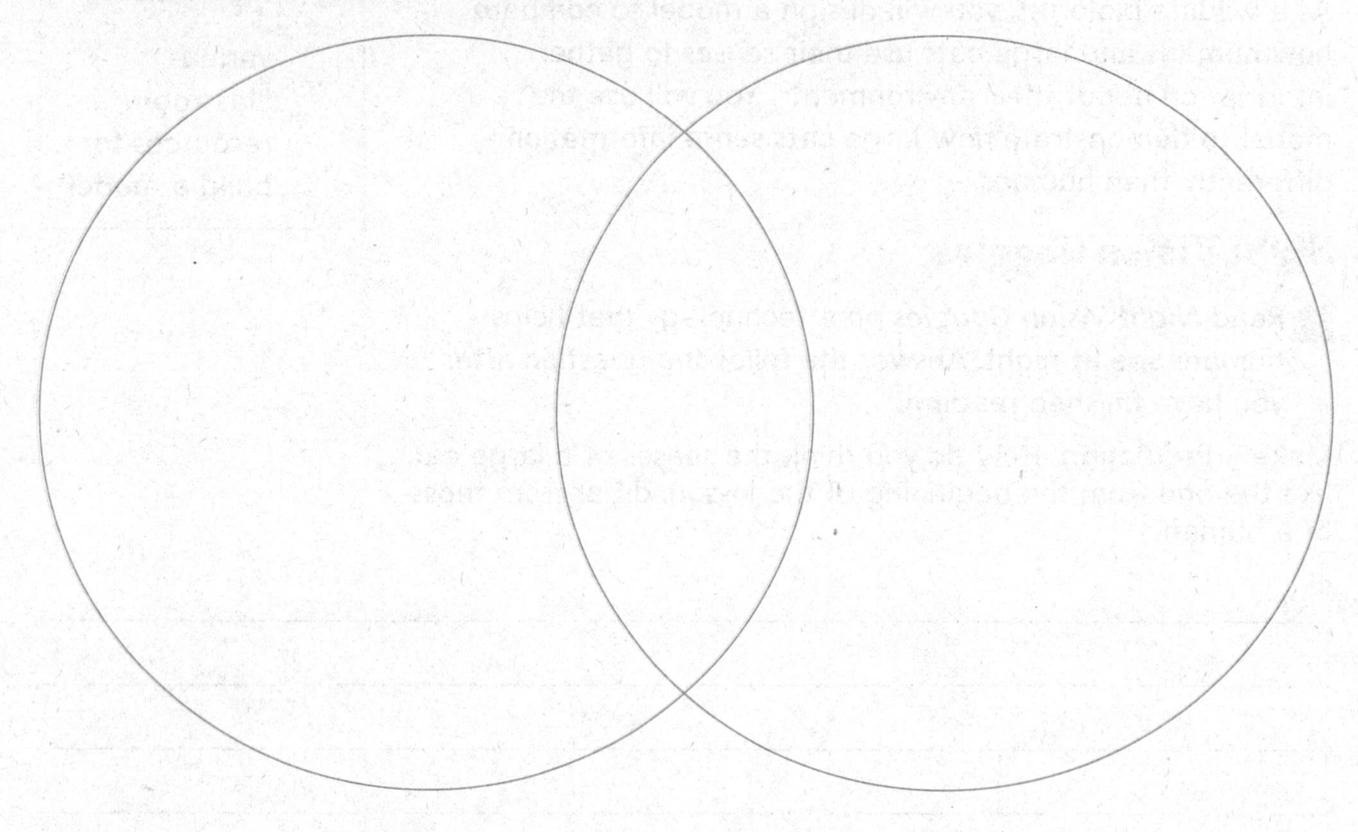

Crosscutting Concepts

Systems and System Models

2. Explain how your model shows the differences between human senses and big cat senses.

Essential Question

How do animals sense and respond to information?

Think about the photo of the large cat hunting at night. Explain how the cat senses and responds to information.

____

____

____

____

____

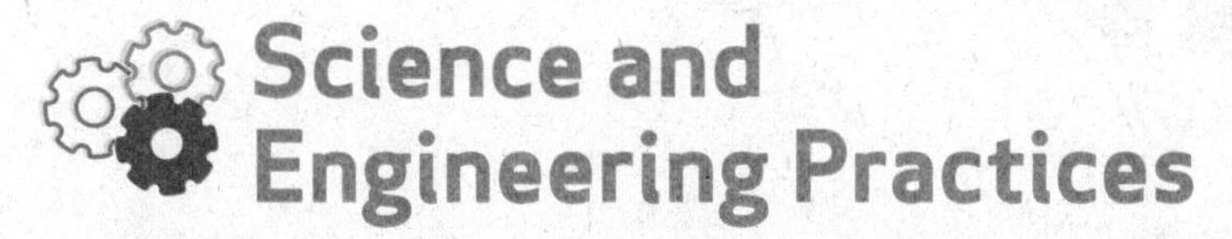

Science and Engineering Practices

Now that you're done with the lesson, share what you did!

Review the "I can . . ." statement you wrote earlier in the lesson. Explain what you have accomplished in this lesson by completing your "I did . . ." statement.

I did ______________________________________

____

____

____

 Name ______________________ Date __________

The Role of Animals' Eyes

Animal Eyes

Two friends were talking about animals' eyes. They noticed that animals that are active at night have very large eyes. This is what they said:

Laura: *Animals like owls have large eyes so they can see in the dark. They do not need light to see.*

Jayden: *Animals like owls have large eyes to help them see in the dark. They still need some light to see.*

Who do you agree with most? ______________________

Explain why you agree.

____

____

____

____

Name ______________________ Date ____________ ENGAGE

Science in Our World

Look at the photos of animals' eyes. What questions do you have?

____

____

____

____

____

Read about a veterinarian and answer the questions on the next page.

STEM Career Connection

Veterinarian

As an animal doctor, I help many different types of animals with medical issues. Sometimes I perform a routine physical exam to make sure the animal is healthy. At other times, I have to perform surgery on animals. I enjoy helping all animals, but horses are my favorite! I made a visit to a farm today to check on the progress of a patient of mine. She's a horse that is having some eye problems.

This horse has an opaque lens. She is suffering from cataracts, which is very similar to the condition observed in humans. Cataracts can cause blindness if not treated. I will recommend surgery to improve this horse's condition.

Treating eye problems in animals is an important job. Veterinarians need to understand the structures and functions of animals' eyes to provide effective care.

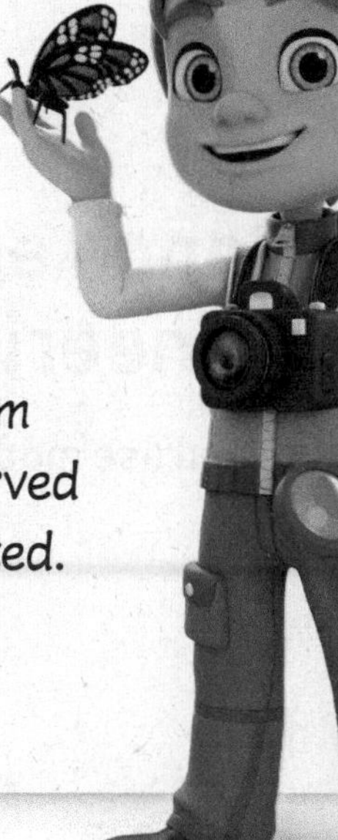

OWEN
Entomologist

Name ______________________ Date __________

1. How is a veterinarian similar to a doctor you might go to?

2. What might happen to the horse if its medical condition is not treated?

Essential Question
How do animals see?

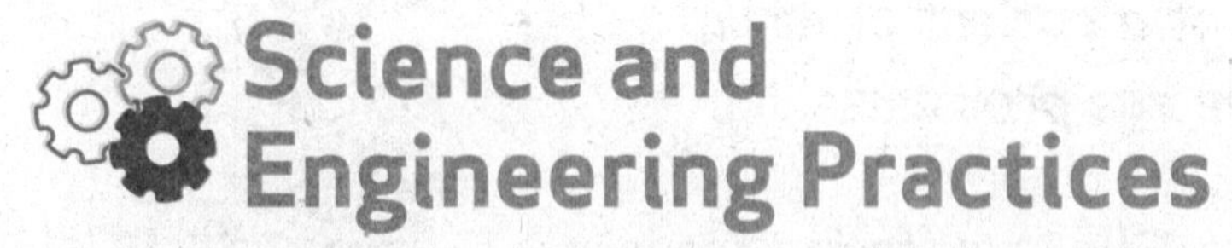

Science and Engineering Practices

I will develop and use models.

Name ______________________ Date ____________ EXPLORE

Inquiry Activity
In the Blink of an Eye

Materials

- [ ] ***Science Handbook***
- [ ] stopwatch
- [ ] bright light source
- [ ] electric fan

What makes animals, like humans, blink? In this activity you will conduct an investigation to explain why you think we blink. You will change several conditions and see how that affects the rate of blinking.

Make a Prediction How does changing the amount of light or wind in an environment affect the rate of blinking in animals?

____

____

____

Carry Out an Investigation

BE CAREFUL When using the fan, make sure there is no loose material that will blow into a person's face.

1. You will work with a partner to do this activity. Have your partner begin reading from the ***Science Handbook*** or other reading material.

2. **Record Data** Once your partner is reading, start the stopwatch. For one minute, you will count the number of times your partner blinks while reading. Record these data in the table below.

	Number of Blinks per Minute: Your Results	Number of Blinks per Minute: Partner Results
Reading Normally		
In Bright Sunlight		
With Wind		

3. Repeat steps 1 and 2, but this time, have your partner sit in very bright light.
4. Repeat steps 1 and 2, but this time, have your partner sit in front of a fan blowing air gently onto your partner while he or she reads.
5. Switch places with your partner and repeat steps 1-4. Make sure that you and your partner each record data in your own ***Be a Scientist Notebook***.
6. **Analyze Data** Look at the data you collected. What pattern or patterns do you see?

Communicate Information

1. Why do you think there was a change in the rate of blinking?

2. What inference can you make about why other animals blink?

3. Do you think that all animals blink? Explain.

Name ______________________ Date __________

Obtain and Communicate Information

Vocabulary

Use these words when explaining how animals are able to see.

image	reflection	refraction
concave lens	convex lens	transparent
translucent	opaque	

The Way Eyes See It

Read *The Way Eyes See It* on how the eye works to allow animals to see. Answer the following questions after you have finished reading.

1. Using information from the Science File, draw and label a human eye below.

 Name ______________________ Date ________

Crosscutting Concepts
Systems and System Models

2. Explain how light enters the eye, is focused on the retina, and is interpreted by the brain.

____

____

____

____

____

____

____

3. What is the meaning of the word *pupil* on page 2? What is another meaning for the word *pupil*? What clues in the text show you which meaning is used on page 2?

____

____

____

____

____

____

How Do Animals See?

Watch the video *How Do Animals See?* about animal eyesight. Answer the question after you have finished watching.

4. How do dogs see differently than humans?

____

____

____

FOLDABLES®

Cut out the Notebook Foldables tabs given to you by your teacher. Glue the anchor tabs as shown below. Use what you have learned to compare and contrast human eyes and owl eyes.

Glue anchor tab here

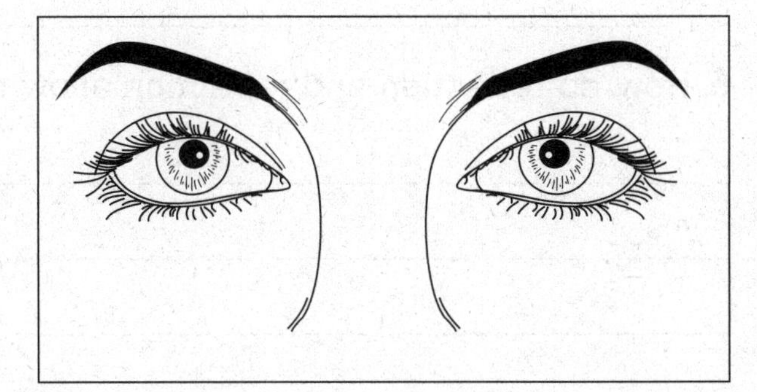

 Name ______ Date ______

Bouncing and Bending of Light

Read pages 340–341 in the ***Science Handbook***.
Answer the following questions after you have finished reading.

5. What is the difference between refraction and reflection?

6. How do reflection and refraction allow animals to see?

7. Explain the difference in how a concave lens and a convex lens affect an object's appearance.

Name ______________________ Date ____________ EXPLAIN

Transparent, Translucent, and Opaque

Read pages 342–343 in the ***Science Handbook***.
Answer the following questions after you have finished reading.

8. What does it mean if an object is transparent? Give an example of when having a transparent material is helpful.

____

____

____

____

____

____

9. How would you know if a material is opaque? When would you find opaque materials useful?

____

____

____

____

Science and Engineering Practices

Use examples from the lesson to explain what you can do!

Think about the model of the human eye that you drew. Tell how you can develop and use models by completing the "I can . . ." statement below.

I can ____

____

____

 Name ______________________ Date ________

Research, Investigate, and Communicate

Why Do Some People Need Eyeglasses?

Explore the Digital Interactive *Why Do Some People Need Eyeglasses?* on reasons why people need corrective lenses for clearer vision. Answer the questions after you have finished.

1. Describe how you would see objects if you were nearsighted.

2. What kind of lens would you need to correct your vision if you were nearsighted? How do you know?

3. What kind of lens would you need to correct your vision if you were farsighted? How do you know?

Name ____________________ Date __________ EVALUATE

Performance Task

It's Time to Focus

Materials

- [ ] hand lens
- [ ] table lamp with clear incandescent bulb, no shade
- [ ] projection screen

As a veterinarian, you will make a model to show how an animal eye works to refract light. You will investigate what happens when you change the distance between the lens and retina in a model eye.

Write a Hypothesis How is the quality of an image affected as you change the distance between the retina and the lens? Write a hypothesis in the form of an "If..., then..." statement.

____

____

____

____

Make a Model

BE CAREFUL Lightbulbs can get very hot!

1. Experiment with using the hand lens and lamp to refract light, forming a crisp image of the lamp on the projection screen. Draw and label a diagram of your setup.

Crosscutting Concepts

Cause and Effect

1. What did you have to do to get a crisp image of the lamp on the screen?

______

______

______

Cause and Effect

2. What did you notice about the image you focused on the screen? Why did the image appear that way?

______

______

______

______

Systems and System Models

3. How is your setup a model of how the eye works in animals?

______

______

______

4. Does the result of your investigation support your hypothesis? Explain.

Essential Question

How do animals see?

Think about the photo of the animals' eyes you saw at the beginning of the lesson. Using what you have learned in this lesson, explain how the animals in that photograph are able to see.

Science and Engineering Practices

Review the "I can . . ." statement you wrote earlier in the lesson. Explain what you have accomplished in this lesson by completing your "I did . . ." statement.

I did ________________

 Name ______________________ Date __________

Structures and Functions of Living Things

Performance Project

How are the shapes and sizes of animals' eyes related to their functions?

Look back at the questions that you wrote about the cuttlefish's eye, and at the work of the marine biologist. You have learned about the structures and functions of living things and can probably answer many of those questions. Marine biologists study organisms that live in the ocean. They observe how organisms behave and interact with the environment. The field notes that you read were taken by a marine biologist that studies animals that live deep in the ocean where there is little to no light. She makes observations, and takes photos and measurements to document and attempt to explain the structures of the animals living in this environment. Now it is your turn to investigate how structures affect function in animals' eyes.

Using models, you will show how the size of an animal's pupil affects how it sees. Use paper plates to model four different-sized pupils of animals. Poke a different-sized hole, starting with a pinhole, in each paper plate to model each pupil. After constructing your models, look through each model pupil and observe the same object in the distance. Draw and label each model eye and describe the quality of the image you observed.

Name ______________________ Date ____________

How are the shapes and sizes of animals' eyes related to their functions?

Eyes differ from one animal to the next. Some eyes are large, and some are small. Some have large pupils, and some have small pupils. Some have pupils that are not round. Use what you have learned throughout the module to develop an argument. Explain how the structure of an animal's eye can affect its ability to see at night.

____

____

____

____

____

____

Explore More in Our World

Did you learn the answers to all of your questions from the beginning of the module? If not, how could you design an experiment or conduct research to help answer them?

 Name ________________ Date ________

Wave Patterns and Information Transfer

Science in Our World

Look at the photo of children using a communication device. What questions do you have about what is happening in the photo?

____

____

____

____

____

____

____

____

____

abc Key Vocabulary

Look and listen for these words as you learn about wave patterns and information transfer.

amplitude	binary code	echo
echolocation	frequency	medium
pitch	sound wave	volume
wavelength		

How does information travel to and from the devices we use to communicate?

DEVEN
Sound Engineer

STEM Career Connection

Telecommunications Engineer

When I was a kid, I knew that I wanted to become an engineer because I liked to fix things and solve problems. Now, as a telecommunications engineer, that is what I do every day. I design solutions to communications problems. I work mostly with Internet, computer, and cell phone technologies. This field is changing every day as new technology is invented. I keep up with the latest information in my field so I can solve problems in new and better ways. How do I do this? I read scientific publications and attend conferences with other engineers and scientists.

Draw and label a diagram to show how you think information travels from its source to your cell phone or computer.

Science and Engineering Practices

I will develop and use models.

I will construct explanations and design solutions.

 Name ______________________ Date ____________

How Waves Move

Waves and Matter

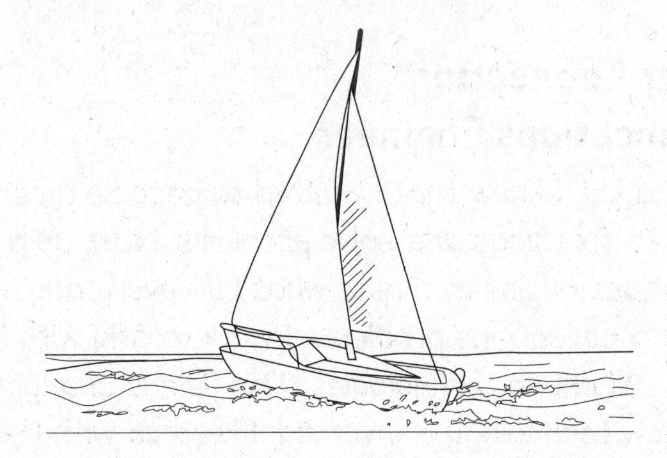

Four friends were talking about waves and their patterns of motion. They agreed that sound waves move through air and water waves move through water. They each had different ideas about whether waves carry matter along with them.

Trisha: *Both types of waves move through matter, but the matter is not carried from one place to another by the sound or water waves.*

Gordon: *Both types of waves move through matter, but only the water waves carry matter from one place to another.*

Otto: *Both types of waves move through matter, but only sound waves carry matter from one place to another.*

Juanita: *Both types of waves move through matter, and both carry matter from one place to another.*

Who do you agree with the most? ______________________

Explain why you agree.

____

____

____

____

Science in Our World

Watch the video of waves. What questions do you have?

____

____

____

____

____

____

Read about an ultrasound technician, and answer the questions on the next page.

Sound waves can help doctors answer questions and solve problems. An ultrasound technician uses sound wave technology to help patients.

STEM Career Connection

Ultrasound Technician

I cannot imagine a job that is more rewarding than mine! As an ultrasound technician, I help diagnose medical conditions by creating images of the inside of the human body. I do this using a technology called ultrasound, which uses sound waves to "see" inside the body. Ultrasound is also used to monitor the growth and development of babies before they are born. It is so exciting to show parents the first picture of their baby! Much of my job involves using computers and operating the ultrasound equipment. I also apply my knowledge of human anatomy to help my patients understand their diagnoses.

GRACE
Computer Engineer

 Name ______________________ Date __________

1. How does sound help ultrasound technicians do their job?

____

____

2. What things do ultrasound technicians need to understand to be able to do their job?

____

____

____

____

Essential Question

How do waves travel?

____

____

____

____

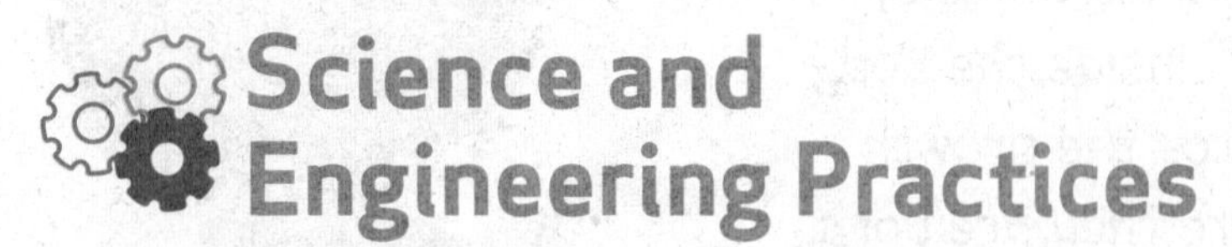

Science and Engineering Practices

I will develop and use models.

Like an ultrasound technician you will understand how waves move by using models.

Inquiry Activity
What Makes Sound?

What causes the sounds we hear around us?

Materials

- [ ] goggles
- [ ] toothpick
- [ ] paper cup
- [ ] scissors
- [ ] rubber band
- [ ] plastic ruler
- [ ] masking tape

Make a Prediction You will conduct an investigation to discover what is needed to make sound. You will also explore ways to change a sound. What do you think will happen to the sound that a rubber band makes when you change the amount of force that is used to pluck the rubber band?

____

____

____

____

____

____

Carry Out an Investigation

BE CAREFUL Wear goggles. Rubber bands can break and fly into someone's eyes if they are plucked or pulled with too much force. Toothpicks can be sharp. Use caution when handling toothpicks.

1. Use the toothpick to poke a small hole in the bottom of the paper cup.
2. Use the scissors to cut the rubber band. Tie one end of the rubber band to the toothpick.
3. Thread the rubber band through the hole in the paper cup so that the toothpick is inside the cup. The toothpick should hold the rubber band in place as you pull on the rubber band from the other side of the cup.
4. Place the cup upside down on a desk or table. Place the ruler so that it is standing upright against the cup. Use masking tape to tape the ruler to the cup.
5. Stretch the rubber band and tape it to the top of the ruler.
6. **Record Data** Keep the cup and ruler on the desk. Hold the cup in place with one hand and then pluck the rubber band. Record your observations in the table on the next page.

7 Experiment with using a small amount and a larger amount of force to pluck the rubber band. Observe how the sound produced is affected. Record your observations in the table.

Strength of pluck	Observations
First pluck	
Second pluck (less force than first pluck)	
Third pluck (more force than first pluck)	

Communicate Information

Crosscutting Concepts

Cause and Effect

1. How did the sound your device made change as you changed the amount of force used to pluck the rubber band?

2. Use your observations from the activity to form an explanation about how you think your device made sound.

3. How does your prediction compare with the results of this activity?

 Name ______________________ Date __________

Obtain and Communicate Information

Vocabulary

Use these words when explaining how waves travel.

sound wave	medium	wavelength
frequency	pitch	amplitude
volume		

Sound Waves and How They Move

Read pages 328–330 in the ***Science Handbook.*** Answer the following questions after you have finished reading.

1. Explain in your own words what a sound wave is.

____

____

____

Crosscutting Concepts

Patterns

2. Draw three illustrations showing each type of medium. Compare how sound travels through each medium.

____

____

____

Name ______________________ Date __________ EXPLAIN

Solid	Liquid	Gas

3. Explain why sound waves cannot travel in space.

____

____

Inquiry Activity
Sound Carriers

Materials

- ☐ small radio
- ☐ wooden desk or table
- ☐ plastic storage bag full of water

You will conduct an investigation to see how sound travels through each state of matter. You will record observations as you conduct your investigation.

Write a Hypothesis Which medium do you think you will be able to hear sound through best: air, water, or wood? Record your hypothesis in the form of an "If. . ., then. . ." statement.

____

____

____

____

Name ______________________ Date ____________

	Volume of Sound Through Medium
Bag of water	
Wooden table	
Air	

Carry Out an Investigation

1 **Record Data** Turn on the small radio and adjust the volume so you can hear it. Hold the plastic bag full of water against your ear, and place the radio on the other side of the bag and listen. Move your ear away from the bag and listen. Turn off the radio and record your observations.

2 Place the radio on a wooden table. Turn on the radio. Put your ear on the table the same distance away from the radio as your ear was when you listened through the bag of water.

3 **Record Data** Lift your head from the table and listen to the radio. Record your observations for steps 2 and 3 in your table.

4 **Analyze Data** Rank air, wood, and water from best to worst in terms of how they transferred sound.

____

Communicate Information

4. Did the results of your investigation support your hypothesis? Why or why not?

____

____

____

____

Crosscutting Concepts

Patterns

5. What pattern did you observe about sound transfer through different mediums?

Pitch and Volume

Read pages 332–334 in the ***Science Handbook.***

6. How are frequency and pitch related?

7. Draw an example of a sound wave with a high frequency and a sound wave with a low frequency. Label wavelength, amplitude, crest, and trough on your drawings. How would each wave sound?

8. Explain how amplitude and volume are related.

FOLDABLES®

Cut out the Notebook Foldables tabs given to you by your teacher. Glue the anchor tabs as shown below. Use what you have learned to explain the terms based on the picture of students making music.

Glue anchor tab here.

Science and Engineering Practices

Think about the model you created in the *Sound Carriers* investigation. Tell how you can develop and use models by completing the "I can . . ." statement below.

I can ______________________________

____

____

ELABORATE Name ______________________ Date ____________

Research, Investigate, and Communicate

Seismic Waves

Read pages 180 and 182 in the ***Science Handbook***.

1. What is a seismic wave?

____

____

2. Think about how a wave would look if it were recorded during an earthquake. Draw a picture to show what that wave would look like.

3. Describe how sound waves and seismic waves are similar and how they are different.

____

____

____

____

____

____

____

Name ______________________ Date ____________ EVALUATE

Performance Task
Making Waves

Materials

- ☐ rectangular baking tray
- ☐ water
- ☐ cork

You will use a model to show patterns of amplitude and wavelength. You will describe how the wave causes objects to move.

Write a Hypothesis Using what you have learned, form a hypothesis to describe how you can affect the amplitude and wavelength of a wave. Use the "If. . ., then. . ." statement for your hypothesis.

____

____

____

____

Make a Model

1. Fill the pan halfway with water. Gently move the pan back and forth to create a series of waves.
2. Create these three waves: short wavelength with low amplitude, long wavelength with high amplitude, and short wavelength with high amplitude.
3. **Record Data** Draw an example of each wave that you created. Describe how you changed the wavelength and amplitude of each wave.

4 **Record Data** Place the cork in the water in the middle of the pan and record your observations of the motion of the cork as you create different waves.

____

____

5 Stop moving the pan and allow the cork to stop moving.

6 **Record Data** Observe and record the position of the cork compared to where it started in the middle of the pan.

____

Communicate Information

1. Look back at your hypothesis. Did this activity support your hypothesis? Explain.

____

____

____

____

____

2. How did the waves affect the motion of the cork?

____

____

3. What does this investigation tell you about how waves move?

____

____

____

Name ______________________ Date ____________ EVALUATE

Essential Question

How do waves travel?

Think about the video of the different waves you watched at the beginning of the lesson. Explain how those waves travel from one point to another.

___

___

___

___

___

___

___

___

___

Science and Engineering Practices

Now that you're done with the lesson, share what you did!

Review the "I can . . ." statement you wrote earlier in the lesson. Explain what you have accomplished in this lesson by completing the "I did . . ." statement.

I did ______________________

___

___

___

___

___

Name ______________________ Date __________

How Waves Transmit Information

Moving Information

Caleb and his friends wondered how information is moved from one computer to another. They each had different ideas. This is what they said:

Caleb: *I think the information travels as tiny bits of matter.*

Suki: *I think the information travels as waves.*

Rob: *I think the information travels as rays.*

Pita: *I think the information travels as electricity.*

Jean: *I think the information travels as sound.*

Who do you think has the best idea? ______________________

Explain your thinking.

____

____

____

____

Science in Our World

Watch the video of the computer code. What questions do you have?

____

____

____

Read about a computer programmer and answer the questions on the next page.

While that computer code may look like a strange language to you, computer programmers use it to make computers do some really cool stuff!

STEM Career Connection

Computer Programmer

As a computer programmer, it is my job to write the instructions that a computer needs in order to complete a task. Some tasks are pretty easy, like turning a light on or off. Other tasks are much more complex, such as tracking weather and making predictions. We generally refer to these written instructions as computer code.

Today, I am working on some computer code that will be used by beginner programmers.

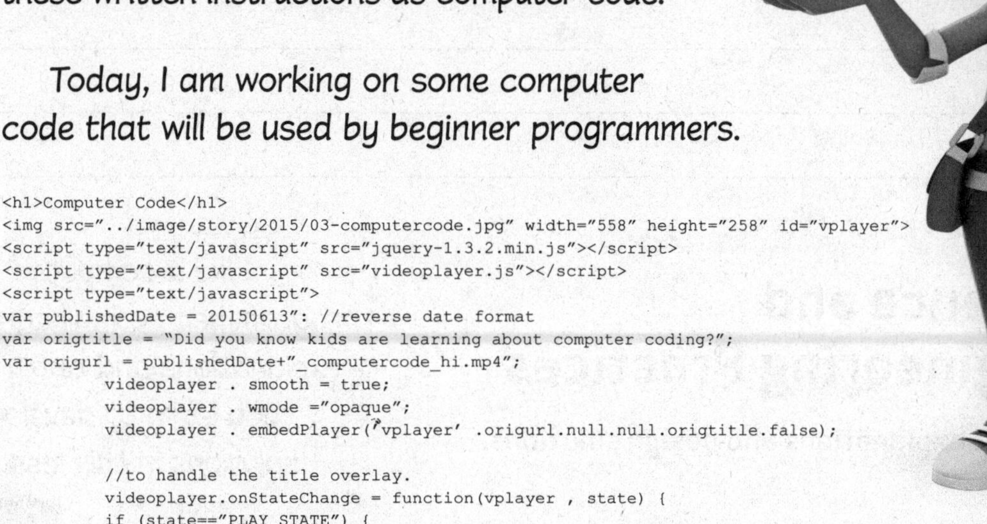

```
<h1>Computer Code</h1>
<img src="../image/story/2015/03-computercode.jpg" width="558" height="258" id="vplayer">
<script type="text/javascript" src="jquery-1.3.2.min.js"></script>
<script type="text/javascript" src="videoplayer.js"></script>
<script type="text/javascript">
var publishedDate = 20150613": //reverse date format
var origtitle = "Did you know kids are learning about computer coding?";
var origurl = publishedDate+"_computercode_hi.mp4";
          videoplayer . smooth = true;
          videoplayer . wmode ="opaque";
          videoplayer . embedPlayer('vplayer' .origurl.null.null.origtitle.false);

          //to handle the title overlay.
          videoplayer.onStateChange = function(vplayer , state) {
          if (state=="PLAY_STATE") {
                $("#kiosk h1").fadeOut ():
          } else if (state=="STOPPED_STATE" || state=="PAUSE_STATE") {
                $("#kiosk h1").fadeIn ();
          }
}
</script>
```

GRACE
Computer Engineer

Name ______________________ Date __________

1. What does a computer programmer do?

---

2. Looking at the computer code, what clues can you find that tell what it might be for?

Essential Question

How do we use patterns and waves to transmit information?

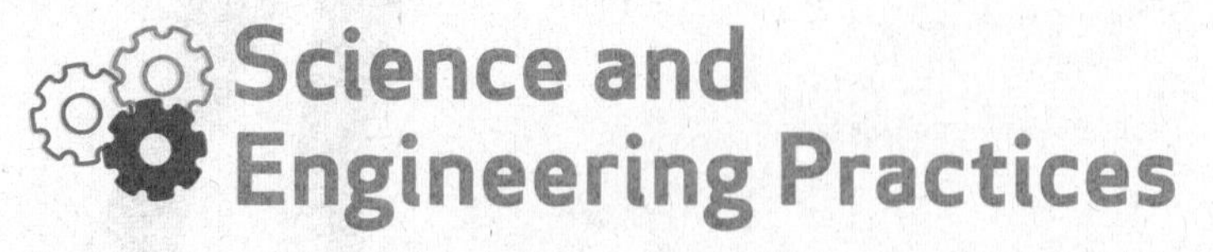

Science and Engineering Practices

I will construct explanations and design solutions.

Inquiry Activity

Using Waves to Transmit Information

How can sound be used to transmit information about location?

Make a Prediction Can you locate the source of sound without using your sense of sight? Explain.

____

____

____

Materials

- [ ] bell or noise-maker
- [ ] material for individual blindfolds
- [ ] measuring tape

Carry Out an Investigation

BE CAREFUL The blindfold should be snug but not uncomfortable. When the blindfold is in place you should not attempt to move around the classroom.

1. Have one classmate sit in a chair. Carefully apply the blindfold to this classmate.
2. With your classmate sitting in the chair blindfolded, ring the bell in front, behind, or to either side of him or her at different distances.
3. Have the seated classmate guess the direction of each ringing sound and the distance from him or her in meters.
4. **Record Data** The third classmate records the actual location of where the bell rang and measures the distance from the bell to the blindfolded classmate using the measuring tape. Record this information in the table on the next page. Repeat steps 2–4 for five trials.
5. Switch roles and repeat steps 1–4 so all team members have a chance to wear their individual blindfolds.

Communicate Information

1. How could you tell where the noise-making classmate was?

____

____

____

 Name ______________________ Date __________

Results for (student name):				
Trial	**Location: Guess**	**Distance (m): Guess**	**Location: Actual**	**Distance (m): Actual**
1				
2				
3				
4				
5				

2. How did you guess the distance the bell was away from you.?

____

____

____

____

____

3. How was your prediction supported by what you discovered in this activity?

____

____

____

____

____

Name ______________________ Date ____________ EXPLAIN

Obtain and Communicate Information

Vocabulary

Use these words when explaining information transfer by waves.

echo echolocation binary code

coding

Information Transfer

Watch *Information Transfer* on ways we send and receive information. Answer the question after you have finished watching.

1. How would you explain information transfer in your own words?

____

____

____

Reflection of Sound

Read page 331 in the ***Science Handbook***. Answer the following questions after you have finished reading.

2. What causes an echo?

____

____

3. Explain how echolocation and sonar are similar.

____

____

____

____

____

____

 Name ______________________ Date __________

Echolocation in Animals

Explore the digital interactive *Echolocation in Animals* on how different animals use sound to find food and to navigate. Answer the question after you have finished.

4. Explain how echolocation works using numbered steps.

____

____

____

____

How Does Technology Help Us Transfer Information?

Read *How Does Technology Help Us Transfer Information?* on ways humans have developed to communicate information. Answer the following questions after you have finished reading.

5. How did humans transfer information before the invention of the telephone?

____

____

6. Compare and contrast how cell phones and satellites transfer information.

____

____

____

____

____

7. How are patterns used in communication technologies?

____

____

____

FOLDABLES®

Cut out the Notebook Foldables tabs given to you by your teacher. Glue the anchor tabs as shown below. Use what you have learned to describe how information can be transmitted digitally and by sound waves.

Glue anchor tab here.

EXPLAIN Name ______________________ Date __________

Inquiry Activity
Morse Code

Materials
- ☐ battery
- ☐ wires
- ☐ lightbulb
- ☐ switch

Make a Prediction How can you send a message using light energy and a simple circuit?

____

____

____

Carry Out an Investigation

1. Work with a partner to build a simple electric circuit with a lightbulb.
2. Think of a short, simple message to send to your partner. Write your message on the next page, then code each letter of your message using dots and dashes from the table below.

Letter	Code	Letter	Code	Letter	Code
A	.-	J	.---	S	...
B	-...	K	-.-	T	-
C	-.-.	L	.-..	U	..-
D	-..	M	--	V	...-
E	.	N	-.	W	.--
F	..-.	O	---	X	-..-
G	--.	P	.--.	Y	-.--
H	....	Q	--.-	Z	--..
I	..	R	.-.		

Your Message: _______________

Your Coded Message:

____

3. Send your message by flashing the light quickly for a dot and long for a dash. Count 3 seconds between each letter and 5 seconds between each word.
4. As you flash the light, your partner will write the pattern in dots and dashes. Using the table, your partner will then decode the message.
5. Switch roles and repeat steps 2–4.

Partner's Coded Message:

____

Partner's Message:

____

Communicate Information

8. What challenges did you face in sending your message?

____

9. In what kinds of situations would Morse code be useful?

____

10. Why don't people use Morse code to communicate today?

____

11. For a message to be sent successfully, what is needed on the receiver's side?

______

______

______

______

______

Science and Engineering Practices

Think about how you have used patterns of waves to send and receive messages. Tell how you can construct explanations by completing the "I can . . ." statement below.

I can ______

______

______

Research, Investigate, and Communicate

What's That Say?

You know that computers use a special language, called binary code, to be able to communicate with other devices.

Research In this activity, you will research binary code.

A computer programmer uses this language, and some other computer languages, to write programs. These programs can be useful and we depend on them to make our lives easier every day!

Take a look at the binary code below. This code represents a message and it will be your quest to decode that message.

Binary Code Message:

01010011	01000101	01001110	01000100	01001001	01001110	01000111	00100000	01001001
01001110	01000110	01001111	01010010	01001101	01000001	01010100	01001001	01001111
01001110	00100000	01010111	01001001	01010100	01001000	00100000	01010111	01000001
01010110	01000101	01010011	00100000	01001001	01010011	00100000	01000011	01001111
01001111	01001100	00100000	01010011	01000011	01001001	01000101	01001110	01000011
01000101								

Recall that you did something similar to this with the *Morse Code* activity. In that activity letters were represented by dots and/or dashes, similar to binary code.

To get started you need to research to discover the binary code for the alphabet in capital letters. You will also need to research the binary code for the space used between words.

Ask a Question What question will you answer with your research?

____

____

____

____

 Name ______________________ Date __________

Carry Out an Investigation

1 Once you have researched and found the needed information, begin decoding the numbers moving from left to right, just as you would read a book. Record the decoded message below. Shade in boxes that are spaces.

Decoded Message:

2 Write the decoded message.

____

____

Communicate Information

1. **Construct an Explanation** Why do you think it is easier to send information using binary code, instead of using actual letters?

____

____

____

____

____

Hearing Echoes

How can you calculate your distance from a surface that reflects an echo?

Time how long it takes between making a sound and hearing its echo. Multiply by the speed of sound. Then divide by 2 since the sound wave makes a two-way trip before you hear its echo!

Suppose it takes 1 second for you to hear your echo after you call down a well. The speed of sound in air is 340 meters per second (m/s). How far are you from the bottom of the well?

seconds × speed of sound ÷ 2 = distance

1 second × 340 m/s ÷ 2 = 170 meters

You are 170 meters from the bottom of the well.

Writing a Number Sentence

Read the problem carefully. What do you know? (The time was 1 second. The speed of sound is 340 meters per second. The sound makes a two-way trip.) What do you need to find out? (the distance)

Decide if you need to use addition, subtraction, multiplication, or division.

Write a number sentence and solve it.

$1 \text{ s} \times 340 \text{ m/s} \div 2 = x$

$x = 170 \text{ m}$

Check to see that your answer makes sense.

Solve each problem in the box provided:

2. You shout in a canyon. Your echo returns 3 seconds later. How far away is the canyon wall?

3. In the ocean, sound travels at 1,500 meters per second. A ship's sonar signal returns in 4 seconds. How far away is the ocean floor?

Performance Task
Pixel Message

Using what you have learned about information transfer, you will work in a group to design a solution to an information transfer problem.

You will design a device or method to send a message of 0s and 1s to your team members on the other side of the classroom. The 0s and 1s represent black and white pixels that form an image (a symbol, letter, or number) in a grid. Your team members on the other side of the room must fill in a 6-by-6 grid with the 0s and 1s and color them in with black or white to decode your message.

Materials

- ☐ bell or noise-maker
- ☐ battery
- ☐ wires
- ☐ lightbulb
- ☐ switch
- ☐ 2 clear plastic pieces of different colors

Define a Problem What must your device do?

Design a Solution

1. Record the materials and method you will use to send your message.

2. Using 0s and 1s, record your message in the grid on the next page. Shade your 0s or 1s so you can clearly see your message. Do not tell your team members, who will be receiving the message, what the message is. They will fill in the grid with the message as they receive it, starting in the upper left corner and going from left to right.

3 **Record Data** Using your solution, try to send the message to your team members. Record your results and any problems that you encountered.

4 Adjust your technique so that you can send the message more clearly. Record any adjustments that you make.

5 Switch places with your team members and repeat steps 1–4. Record and decode the second message in the grid below.

Communicate Information

Crosscutting Concepts

Patterns

1. How did you use waves and patterns to send your message?

____

____

2. How did you solve the problem that you identified?

____

____

____

____

Name ______________________ Date ____________ EVALUATE

Essential Question

How do we use patterns and waves to transmit information?

Think about the video about binary code you watched at the beginning of the lesson. Explain how we use patterns and waves to transfer messages in the binary code.

____

____

____

____

____

Now that you're done with the lesson, share what you did!

Science and Engineering Practices

Review the "I can . . ." statement you wrote earlier in the lesson. Explain what you have accomplished in this lesson by completing the "I did . . ." statement.

I did ____

____

____

____

____

 Name ______________________ Date ________

Wave Patterns and Information Transfer

Performance Project

Let's Communicate!

Think back to the journal entry of the telecommunications engineer from the beginning of the module. Telecommunications engineers design the technology and systems that allow us to communicate. Throughout the module, you have learned about different ways messages can be sent. You have also learned about how sound travels. Now it is your turn to be a telecommunications engineer and design a device to send a spoken message using sound waves.

You will use the materials your teacher provides to design a communication device that can send a message up to 10 meters (m). Use what you have learned throughout the module to choose the best materials for your design. Test your device and try different ways to make sound travel through your design. Draw a picture of your design and label the materials that you used.

How does information travel to and from the devices we use to communicate?

Try talking to your classmate through the device that you designed. Record your observations and revise your design as needed. If you are unable to transmit your message by voice, then develop a code for communicating using your device. Which method and materials worked best to send a spoken or coded message?

Explore More in Our World

Did you learn the answers to all of your questions from the beginning of the module? If not, how could you conduct research or design an experiment to help answer them?

 Name ______________________ Date __________

Patterns of Earth's Changing Features

Science in Our World

Look at the photo of the Grand Canyon. Notice the different colors and landforms. What questions do you have?

____

____

____

____

____

____

Key Vocabulary

Look and listen for these words as you learn about Earth's changing features.

deposition	erosion	fossil
landform	landslide	sediment
sedimentary rock	topographical map	vegetation
weathering		

Name ______________________ Date __________

What evidence is used to learn about how Earth's surface has changed over time?

JIN
Paleontologist

STEM Career Connection

Park Ranger

As a park ranger at Grand Canyon National Park, it is my job to preserve and protect the park and its natural resources and to educate the public about the park. This includes helping them understand and appreciate the geology, ecology, and natural history of the park. A big part of my job is answering questions from visitors. Everyone wants to know how the Grand Canyon was formed. I also get a lot of questions about the colors of the rocks and why there seem to be layers on the sides of the canyon. To answer these questions, I help the visitors understand how rocks are formed, how they can break up, and how they can be carried away by water.

Draw and label a diagram to show how you think the Grand Canyon formed.

Science and Engineering Practices

I will carry out investigations.

I will analyze and interpret data.

I will construct explanations.

 Name ______________________ Date __________

Earth's Landforms and Features

Land and Water Features

Four friends were talking about land and water features such as mountain ranges, volcanoes, ocean trenches, and formations on the ocean floor. They wondered which features occur in patterns. This is what they said:

Woojin: *I think the features on land are the ones that occur in patterns.*

Noah: *I think the features found in oceans are the ones that occur in patterns.*

Abby: *I think both the land and ocean features occur in patterns.*

Elena: *I think land and ocean features occur anywhere. They don't follow any patterns.*

Who do you agree with most? ______________________

Explain why you agree.

____

____

____

Science in Our World

Look at the aerial photo of Mount Everest.
What questions do you have about the image?

______
______
______
______
______
______

Read the cartographer's field notes and answer the questions on the next page.

STEM Career Connection

Cartographer

Questions: How is the construction in town changing the land? How are changes to the land affecting the habitats of aquatic organisms?

Spatial Data Needed: land-use data, water-quality data, aquatic-biota data

Scientists and Engineers to Contact for Data: civil engineer, soil scientist, aquatic ecologist

Plan: Use geographic information systems to create a map showing how the location of all of the data are related. Track changes in the data over time as construction and development occur.

Name ______________________ Date __________

1. What do you think the cartographer does based on the field notes?

___

___

___

2. What do you think a geographic information system is?

___

___

___

Essential Question

What are Earth's features?

___

___

___

___

___

Science and Engineering Practices

I will analyze and interpret data.

Like a cartographer, you will analyze and interpret data.

Name ______________________ Date ____________

Inquiry Activity

Moving Earth

How can the ground move during an earthquake?

Make a Prediction What will happen to the the ground in a model earthquake?

____

____

Materials

- [ ] cut pieces of foam
- [ ] aluminum pan
- [ ] soil
- [ ] wooden block

Carry Out an Investigation

1. Place the two pieces of foam in a pan so the cut surfaces touch each other.
2. Cover the foam with soil, and smooth the soil over both pieces of foam.
3. Pull about 5 centimeters of the pan off the edge of a table.
4. Gently tap the bottom of the pan with a block.
5. **Record Data** What happened to the foam and the soil?

____

____

____

____

____

____

6. What happens as you continue to tap the pan?

____

____

____

____

____

____

Name ______________________ Date __________

1. **Analyze Data** What do you think would happen if you tapped the pan harder?

2. What do you think the foam blocks and the cut between the blocks represent?

3. How could an event such as this affect Earth's features?

Obtain and Communicate Information

Vocabulary

Use these words when describing Earth's landforms and features.

landform	continent	plate tectonics
volcano	earthquake	fault
topographical map		

Landforms

Watch *Landforms* on the features that cover Earth's surface. Answer the following questions after you have finished watching.

1. What types of landforms did you see in the video?

____

____

____

2. What types of landforms have you seen?

____

____

____

Types of Landforms

Read pages 141–143 in the ***Science Handbook***. Answer the following questions after you have finished reading.

3. What is a landform?

____

____

4. What is a valley?

5. What is a plateau?

6. What is a desert?

7. What is a tributary?

Name ____________ Date ____________ EXPLAIN

FOLDABLES®

Cut out the Notebook Foldables tabs given to you by your teacher. Glue the anchor tabs as shown below. Use what you have learned to define the landforms shown.

Glue anchor tab here.

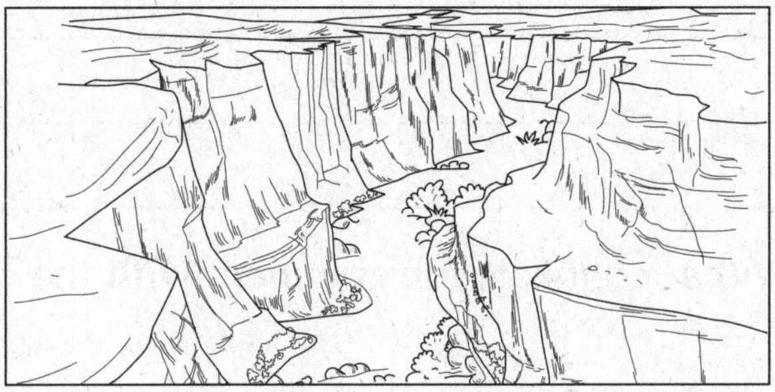

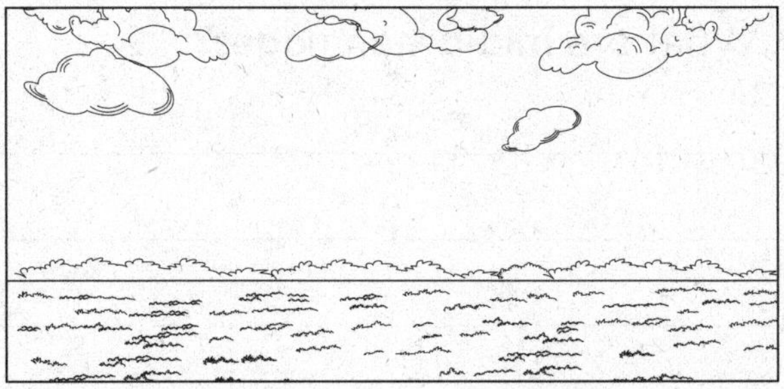

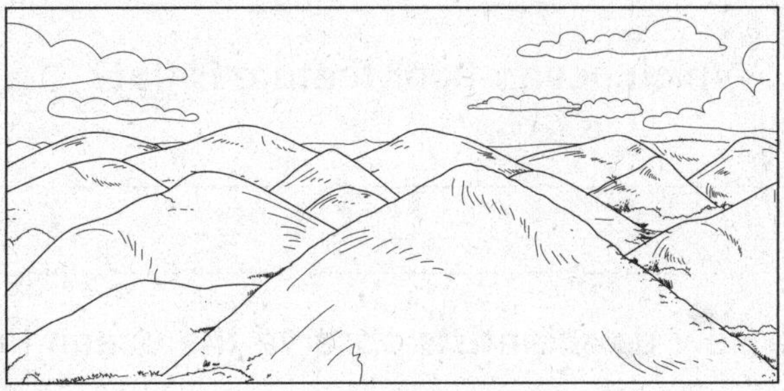

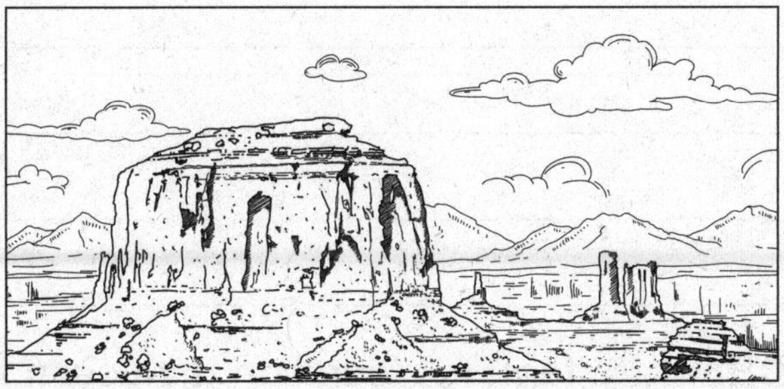

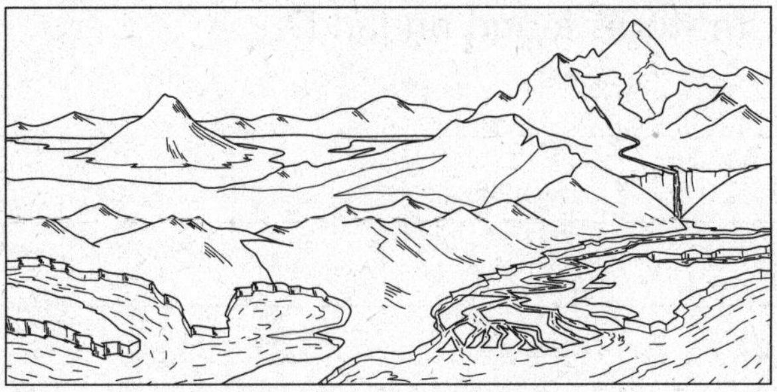

EXPLAIN Name ____________________ Date __________

Earth's Ocean Features

Read pages 144–145 in the ***Science Handbook***. Answer the following questions after you have finished reading.

8. How are the continental shelf and continental slope related?

9. What connects the continent with the ocean floor?

10. What are mid-ocean ridges?

11. Which ocean-floor feature is flat?

12. How do scientists observe the ocean floor?

13. How are features found on the ocean floor similar to features found on land?

Name ______________________ Date __________ EXPLAIN

How Earthquakes and Volcanoes Shape Earth

Read *How Earthquakes and Volcanoes Shape Earth*. Answer the following questions after you have finished reading.

14. What happened 250 million years ago?

15. How do colliding plates form a volcano?

16. What is a fault?

World Earthquakes and Volcanoes

Investigate how the locations of tectonic plates are related to the locations of volcanoes and earthquakes by conducting the simulation. Answer the questions after you have finished.

17. What do you observe about the plate boundaries?

18. Look at the arrows. Why do you think earthquakes happen when plates move?

How Scientists Use Maps

Read page 418 in the ***Science Handbook***. Answer the following questions after you have finished reading.

19. Why do people use maps?

20. What types of information might be featured on a map?

21. What does a relief map show?

Patterns on Earth's Surface

Explore the Digital Interactive *Patterns on Earth's Surface*. Answer the question after you have finished.

22. What do you notice about the locations of earthquakes and volcanoes?

Science and Engineering Practices

Think about the evidence you have seen about Earth's features. Express how you can analyze and interpret data by completing the "I can..." statement below.

I can ______

Name ______________________ Date ____________ ELABORATE

Research, Investigate, and Communicate

Topographic Maps

Read the third paragraph on page 415 in the ***Science Handbook***. Answer the following questions after you have finished reading.

1. How are the contour lines arranged in an area with a very gradual slope?

____

____

____

____

2. How are the contour lines arranged in an area with a very steep slope?

____

____

____

____

3. How would the lines be arranged on a topographical map to show where a mountain was located?

____

____

____

____

____

Performance Task

Landforms from Another Planet

Materials
☐ soft modeling clay
☐ pencil
☐ ruler
☐ dental floss
☐ graph paper

Suppose that a distant planet has just been discovered. The planet is very similar to Earth. Information, gathered by deep-space telescopes, shows that the planet contains several common landforms that are found on Earth. Data have also shown that this newly discovered planet has oceans of water, mountain ranges, volcanoes, ocean trenches, and ocean ridges. You will create a topographic map of a landform found on the planet.

Ask a Question What question will people on Earth be able to answer about the landform using your topographic map?

____

____

____

____

Make a Model

1. Choose a landform that is found on the newly discovered planet and make a model of it using modeling clay.
2. Using the pencil, poke a hole through the center of your landform.
3. Measure the height of your landform in centimeters.
4. Using the dental floss, cut a slice 1 cm in thickness off the top of the landform.
5. Place the slice of landform onto a piece of graph paper. Trace the edges of the slice and mark the location of the hole in the middle. Then put the slice to the side.

6. Cut the next 1 cm slice of the landform. Place it on the graph paper so the hole in the middle of the slice lines up with the spot you marked on the graph paper from the previous slice. Then trace the edges of the slice.
7. Cut, line up, and trace the slices for the rest of the landform. When you are finished, put the clay landform back together.
8. Switch maps with a classmate. Do not show your classmate the model landform that you created. Have the classmate use your map to try to recreate your model landform with modeling clay.

Glue your graph here.

Communicate Information

1. Explain how you used the topographic map to create your classmate's model landform.

____

____

____

2. Compare your classmate's original model landform to the one that you created with your classmate's map. How close were you to the original model landform?

____

____

____

____

Crosscutting Concepts

Patterns

3. Where would you be likely to find the landform that you mapped on the distant planet? Place an X on the map of the planet below. Use what you know about patterns in the locations of Earth's features to form an explanation about where the landform is likely to be found.

___

___

___

Essential Question

What are Earth's features?

Think about the photo of Mount Everest you saw at the beginning of the lesson. What do you know now about where mountains are located?

___

___

___

___

___

___

___

Name ______________________ Date ____________ EVALUATE

Science and Engineering Practices

Review the "I can . . ." statement you wrote earlier in the lesson. Explain what you have accomplished in this lesson by completing your "I did . . ." statement.

I did ______________________________________

____

____

____

____

____

____

____

Name ______________________ Date __________

Effects of Erosion

Changing Landforms

Three friends were talking about processes that change landforms. They each agreed that wind, water, and ice could wear away and change landforms. They disagreed on how long it takes for these processes to change landforms. This is what they said:

Bianca: *I think it takes a long time for these processes to change landforms.*

Rose: *I think it takes a short time for these processes to change landforms.*

Wyatt: *I think these processes sometimes change landforms quickly, but they can also take a long time.*

Who do you agree with most? ______________________

Explain why you agree ______________________

____

____

____

____

Name ______________________ Date ____________ ENGAGE

Science in Our World

Look at the photo of a landslide. What questions do you have about what is happening in the picture?

____

____

____

____

Read the soil conservationist's journal entry and answer the questions on the next page.

Soil conservationists work to reduce the effects of the processes that change landforms.

STEM Career Connection

Soil Conservationist

After many years of working with farmers to reduce soil loss from their fields, we are finally making progress. As a soil conservationist, I help farmers make decisions about their land so that it will produce the most crops, while not damaging the soil or water resources. One method used to reduce the loss of soil from fields is to change the types of crops planted by the farmer. The switching of crops being planted is called crop rotation. Crop rotation helps reduce the amount of soil lost by keeping it from being exposed to water and wind.

If wise soil conservation practices are not followed, water running over the land can carry away soil and nutrients. The water and soil then flow into rivers and lakes. Over time, lakes and rivers can start to get filled in by the soil.

Changes to landforms happen constantly. Sometimes these changes are not helpful, like a farmer's field losing the top layer of nutrient-rich soil.

MAYA
GEOLOGIST

 Name ____________________ Date __________

1. What does a soil conservationist do?

2. What could happen if a farm did not use soil conservation methods?

Essential Question
How do living and nonliving things change Earth's surface?

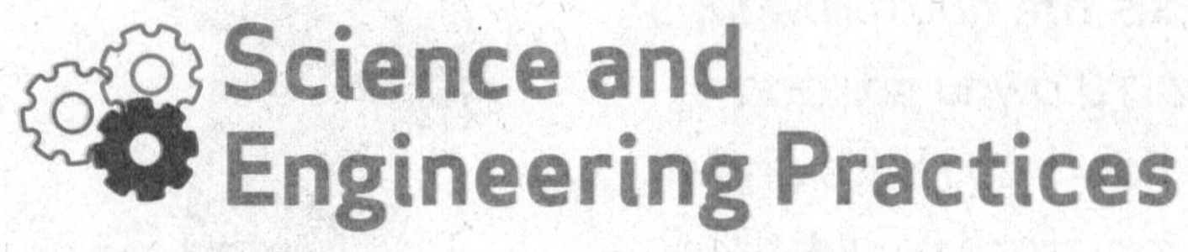

I will plan and carry out investigations.

Name ______________________ Date ____________ EXPLORE

Inquiry Activity
Shake, Rattle, and Roll

How can you model weathering using rocks and moving water?

Make a Prediction What will happen to rocks if you shake them in water?

____

____

____

Materials

- ☐ safety goggles
- ☐ 3 plastic jars with lids
- ☐ sandstone rocks
- ☐ measuring cup
- ☐ water
- ☐ stopwatch
- ☐ hand lens

Carry Out an Investigation

BE CAREFUL Wear safety goggles.

1. Label three jars *No Shake, 2-Minute Shake,* and *5-Minute Shake.* Place the same number of similar-sized rocks in each jar.
2. Using the measuring cup, fill each jar with the same amount of water. Put a lid on each jar.
3. Let the *No Shake* jar sit. Do not shake it.
4. Shake the *2-Minute Shake* jar hard for 2 minutes. Then let the jar sit.
5. Shake the *5-Minute Shake* jar for 5 minutes. The let the jar sit.
6. Use a hand lens to observe the rocks in each jar.

EXPLORE Name ______________________ Date __________

1. **Record Data** What happened to the rocks? Record your observations in the table below.

	Observations
No Shake	
2-Minute Shake	
5-Minute Shake	

2. **Analyze Data** How can rocks change in moving water?

____

____

____

Communicating Information

3. What do you think would happen if you shook the jar for an hour?

____

____

4. How did your results compare with the prediction you made?

____

____

____

Obtain and Communicate Information

Vocabulary

Use these words when describing erosion.

weathering	acid rain	vegetation
erosion	deposition	landslide
avalanche		

Weathering and Erosion

Watch *Weathering and Erosion* on the processes that change Earth's surface. Answer the following questions after you have finished watching.

1. What do you think causes the changes in the landforms and man-made objects you observed in the video?

____

____

2. How do you think the beach in the video will change?

____

____

3. When do erosion and weathering occur?

____

____

4. How can human activities change the way that weathering and erosion affect landforms?

____

____

____

____

Weathering

Read pages 162–165 in the ***Science Handbook***.
Answer the following questions after you have finished reading.

5. Define *weathering*.

____

____

6. List the different types of physical weathering.

____

____

____

Inquiry Activity

Weathered by Vegetation

You will investigate plants growing outside to collect evidence that plants can cause weathering.

Make a Prediction What evidence do you think you will observe that could show how vegetation can cause weathering?

____

____

____

Carry Out an Investigation

1. Following your teacher's directions, go outside and look for where plants are growing in paved areas such as the playground or sidewalks.
2. Look closely at how the plants are growing. Observe the effects the plants are having on the area.
3. **Record Data** Record your observations in the table on the next page. Be sure to make note of the location of where you found the plants.

Plant	Location	Observations
1		
2		
3		

4 **Record Data** Try to locate two more plants growing in different areas. Record your observations in the table.

Communicate Information

7. **Construct an Explanation** Using the data that you collected, explain how vegetation could cause weathering.

____

____

____

8. How did your prediction compare with what you observed?

____

____

____

Erosion and Deposition

Read pages 166–171 in the ***Science Handbook***. Answer the following questions after you have finished reading.

9. What is erosion?

____

____

10. What is deposition?

____

____

 Name ________________ Date ________

Dwindling Mountains

How does the rate of erosion affect a landform?

In the table below, you will see the heights of some mountain peaks in the United States.

Heights of Mountain Peaks			
Mountain	State	Height in meters	Height in feet
Mount McKinley	Alaska	6,194	20,320
Mount Whitney	California	4,417	14,491
Mount Shasta	California	4,317	14,162
Wheeler Peak	Nevada	3,982	13,065

Problem Solving

To find the number of years, you can count backward by 2 from 6,194 m to 6,174 m.

6,192 6,190 6,188 6,186
6,184 6,182 6,180 6,178
6,176 6,174

It would take 10 years for the mountain to become 6,174 meters tall.

Another way to solve this problem is to find the number of meters lost. Then you can divide the difference of meters by 2.

$6{,}194 \text{ m} - 6{,}174 \text{ m} = 20 \text{ m}$

$20 \div 2 = 10$

It would take 10 years for the mountain to become 6,174 meters tall.

Mountains erode by small amounts each year. Suppose Mount McKinley erodes 2 meters each year. How many years would it take for the mountain to be 6,174 meters tall? Follow the steps in the Problem Solving box above.

Analyze Data Use a separate sheet of paper to solve the following problems. If the erosion rate is 1 meter each year, what will be the height of:

11. Mount Shasta in 20 years? ________________

12. Mount Whitney in 15 years? ________________

13. Wheeler Peak in 80 years? ________________

FOLDABLES®

Cut out the Notebook Foldables tabs given to you by your teacher. Glue the anchor tabs as shown below. Use what you have learned to describe how the terms relate to the image of the sea arch.

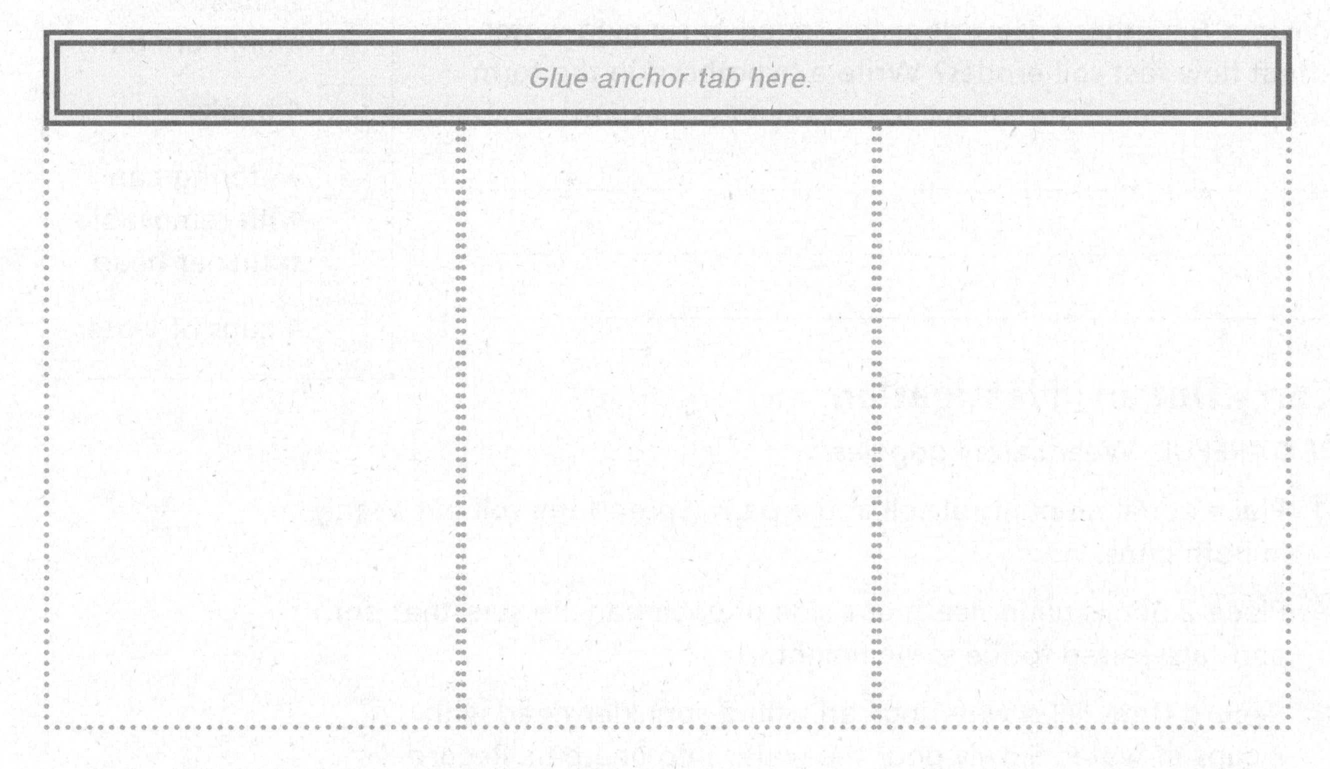

 Name ______________________ Date __________

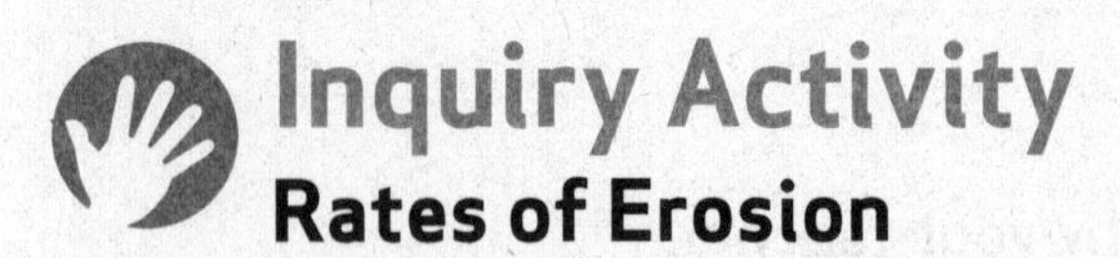

Inquiry Activity
Rates of Erosion

Materials

- ☐ safety goggles
- ☐ 4 cups of soil
- ☐ 2 shallow aluminum pans
- ☐ 4 books
- ☐ watering can with removable sprinkler head
- ☐ 4 cups of water

You will investigate how the speed of moving water affects the rate of erosion.

Write a Hypothesis How does the speed of running water affect how fast soil erodes? Write a hypothesis in the form of an "If..., then..." statement to answer this question.

____

____

____

Carry Out an Investigation

BE CAREFUL Wear safety goggles.

1. Place equal amounts of soil in the pans. Spread the soil out evenly in both pans.
2. Place 2 books underneath one side of each pan. Be sure that both pans are raised to the same height.
3. **Record Data** Fill a watering can with a sprinkler head with 2 cups of water. Slowly pour the water into one pan. Record your observations.

____

____

4. **Record Data** Fill the watering can with 2 cups of water. Remove the sprinkler head. Slowly pour the water into the other pan. Record your observations.

____

____

5. **Analyze Data** How was the rate of erosion different in the two pans?

____

____

____

____

Name ______________________ Date __________

Communicate Information

14. How did your hypothesis compare with the results of the activity?

____

____

____

Crosscutting Concepts

Cause and Effect

15. Do you think the rate of erosion might change in a similar way with wind as the agent of erosion? How? What type of investigation could you plan to test your answer?

____

____

____

____

____

Science and Engineering Practices

Think about what you have learned about weathering and erosion. Express how you can plan and carry out investigations by completing the "I can..." statement below.

I can ______________________

____

____

____

ELABORATE Name ______________________ Date __________

Research, Investigate, and Communicate

Effects of Erosion on Landforms

Investigate the *Effects of Erosion on Landforms* by conducting the simulation. Answer the following question after you have finished.

Communicate Information

1. Choose one of the causes for weathering and erosion. Explain how that process changes different landforms over time. Include in your explanation how that process weathers and then erodes the landforms. After you complete the activity and your explanation, share your response during a class discussion.

Performance Task
Landslide Experiment

You will model two different slopes of loose sediment to observe a landslide.

Write a Hypothesis How does the steepness of a slope affect the downhill movement of soil? Write a hypohesis using an "If..., then..." statement to answer the question.

____

____

____

____

____

Materials

- ☐ safety goggles
- ☐ plastic table-cloth
- ☐ 2 shallow aluminum pans
- ☐ soil
- ☐ rocks of varying size
- ☐ twigs
- ☐ water
- ☐ 6 books
- ☐ 2 baking trays
- ☐ watering can

Carry Out an Investigation

BE CAREFUL Wear safety goggles.

1. Cover your work surface with a plastic tablecloth. Set each pan on the tablecloth.
2. Spread an equal amount soil in the bottom of both pans. Push rocks and twigs into the soil. Use the same amount of soil, rocks, and twigs in each pan.
3. Sprinkle the soil in each pan with water until it is just damp.
4. Prop one end of the first pan up with 1 book.
5. Prop one end of the second pan up with 5 books.
6. Place a cookie tray under the other end of each of the pans.
7. **Record Data** Use a watering can to pour an equal amount of water over the high end of each pan. Record your observations in the table on the next page.

 Name ______________________ Date __________

Pan	Observations
1 book	
5 books	

1. **Analyze Data** Was your hypothesis supported by your observations? Explain.

2. How do you think you could reduce erosion in the pan with the greater slope?

3. Make a plan to test your idea about how you could reduce erosion in the pan with the greater slope.

4. How could you design an experiment to test the effect of another variable, such as the amount of wind or vegetation?

Name ______________________ Date ____________ EVALUATE

Essential Question

How do living and nonliving things change Earth's surface?

Think about the photo of a landslide you saw at the beginning of the lesson. Explain how events like the landslide change Earth's surface.

____

____

____

____

____

____

____

Science and Engineering Practices

Now that you're done with the lesson, share what you did!

Review the "I can . . ." statement you wrote earlier in the lesson. Explain what you have accomplished in this lesson by completing your "I did . . ." statement.

I did ____

____

____

____

____

____

 Name ______________________ Date ____________

History of Earth's Surface

Information from Layers of Rock

Four friends were looking at the layers of rock in a canyon. They each had different ideas about what scientists learn by examining the rock layers. This is what they said:

Anita: *I think scientists examine rock layers to learn about organisms that lived in the past.*

Mason: *I think scientists examine rock layers to learn about how the surface of the Earth changes over time.*

Aaron: *I think scientists examine rock layers to learn about organisms that lived in the past and how the surface of the Earth has changed.*

Owen: *I don't think rock layers help scientists learn about organisms that lived in the past or how the surface of the Earth has changed. They examine rock layers to identify rocks and minerals.*

Who do you agree with most? ______________________

Explain why you agree. ______________________

____

____

Name ______________________ Date __________ ENGAGE

Science in Our World

Look at the photo of the fossils in the layer of rock. What questions do you have?

____

____

____

____

____

____

Read the paleontologist's field notes and answer the questions on the next page.

STEM Career Connection

Paleontologist

Ammonite (Guembelites philostrati)

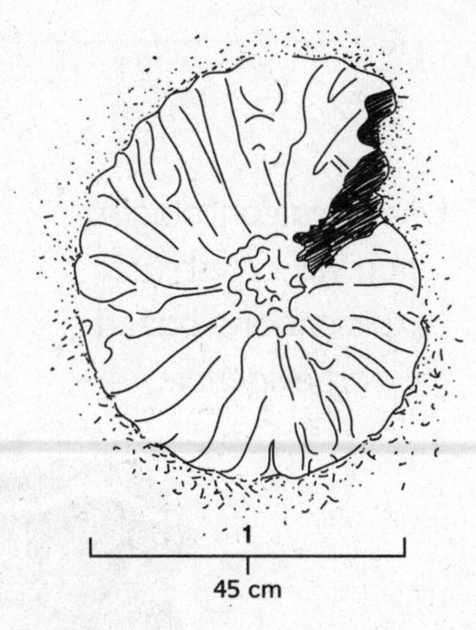

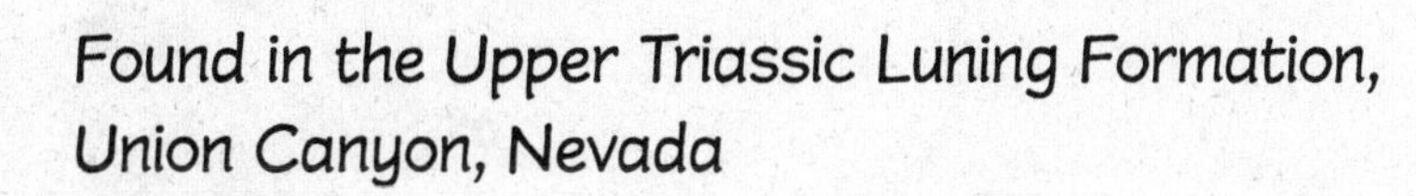

Found in the Upper Triassic Luning Formation, Union Canyon, Nevada

 Name ______________________ Date __________

1. What type of organism do you think an ammonite is? Explain your thinking.

____

____

____

2. In what type of environment do you think ammonites lived?

____

____

____

Essential Question

What can rock formations tell us about Earth's history?

____

____

____

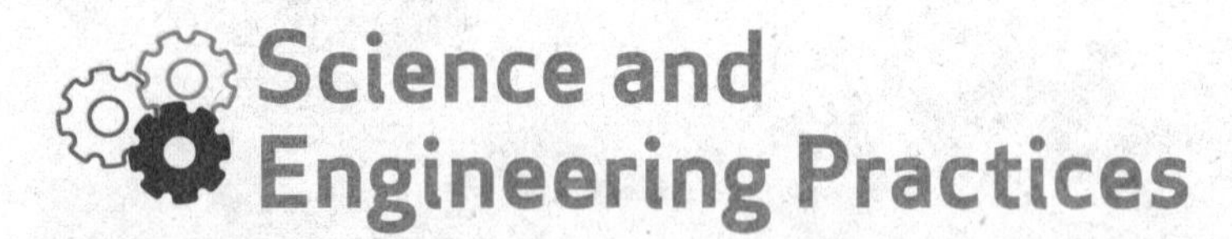

Science and Engineering Practices

I will construct explanations.

Name ______________________ Date ____________ EXPLORE

Inquiry Activity
Making an Impression

What can you tell by looking at footprints?

Make a Prediction You will create a model of fossilized footprints showing an interaction between two animals. Predict what you can infer about the fossil organisms by observing footprints.

____

____

____

Materials

- ☐ modeling clay
- ☐ small objects, such as a pencil, shell, eraser, coin, or paper clip

Carry Out an Investigation

1. Flatten some modeling clay. Use the small objects and your fingers to make imprints by pressing them into the modeling clay. Model different animals-heavy, light, walking, and running.
2. Think of a story about two different animals. Use the objects and your fingers to make model footprints that show your story.
3. Exchange clay models with another group. Try to infer the story that the other group modeled.

Communicate Information

1. How did you model the footprints that different types of animals would make?

____

____

____

____

 Name ____________ Date ____________

2. Explain how you modeled walking and running.

3. How did you model a heavy animal and a light animal?

4. What do you think scientists can infer from footprints made long ago?

5. How did your prediction compare with your observations of the activity?

Name ______________________ Date __________ EXPLAIN

Obtain and Communicate Information

Vocabulary

Use these words when describing Earth's history.

sedimentary rock sediment fossil

index fossil

Sedimentary Rock

Read page 151 in your ***Science Handbook.*** Answer the following questions after you have finished reading.

1. What are sediments?

____

2. How do sedimentary rocks form?

____

____

____

____

____

3. What can relative age tell us about fossils?

____

____

____

____

____

____

EXPLAIN Name ______________________ Date ____________

Fossil Dig

Watch *Fossil Dig* on how paleontologists find and recover fossils. Answer the following questions after you have finished watching.

4. Where can fossils be found in the United States?

____

____

5. What special tools are used to recover fossils?

____

____

6. Why are fossils put into containers or bags?

____

____

7. Why do you think paleontologists are so careful with fossils?

____

____

What Fossils Tell Us

Read pages 214–217 in the ***Science Handbook.*** Answer the following questions after you have finished reading.

8. Which parts of an organism are more likely to leave fossils?

____

____

9. List the types of fossils.

____

____

____

____

10. Explain how scientists can know if dry land was once covered by an ocean.

____

____

____

____

11. What is an index fossil?

____

____

____

____

____

12. How do scientists know that trilobites are older than ammonites?

____

____

____

13. What is extinction?

____

____

____

____

14. How was the fossil of smilodon used?

____

____

____

____

Name ______________________ Date __________

FOLDABLES®

Cut out the Notebook Foldables tabs given to you by your teacher. Glue the anchor tabs as shown below. Use what you have learned to explain how fossils form.

Glue anchor tab here.

Glue anchor tab here.

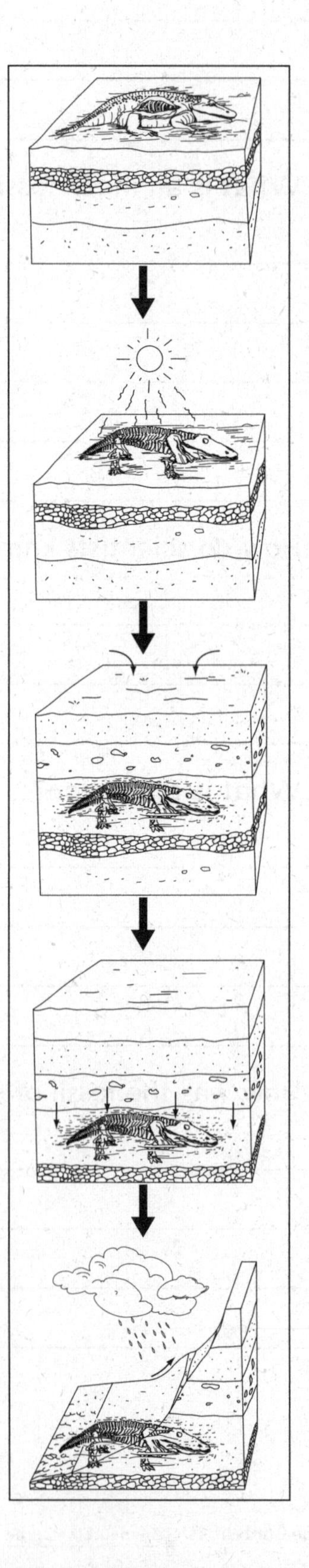

Name ______________________ Date __________ EXPLAIN

Inquiry Activity

Fossil Model

You will create a model that shows fossils and rock layers to explore how scientists learn about the relative ages of fossils.

Make a Prediction How can layers of rock be used to learn about the ages of fossils?

____

____

____

____

Materials

- ☐ 5 different colors of modeling clay
- ☐ small classroom objects, such as erasers and paper clips

1. Flatten each color of modeling clay until each is about ½ inch thick.
2. Place one or two small objects on top of each layer of modeling clay.
3. Place the layers of modeling clay with the objects on them on top of one another to create a stack of modeling clay of different colors. Press the layers down. Switch your model with another classmate. Tell them which side of your model is the top layer. Take note of which color is on the bottom. This is the oldest layer.
4. **Record Data** Draw the layers and label the colors.

5. Carefully pull the layers apart and remove the objects. Record which objects you found in which layers.

____

____

____

Communicate Information

15. Analyze Data Which fossils are oldest and which are youngest? Explain how you know.

16. Does knowing which fossils are older and younger tell you exactly how old they are in years? Why not?

17. How does your prediction compare with the results of the activity?

Science and Engineering Practices

Think about the evidence you have seen about Earth's features. Express how you can analyze and interpret data by completing the "I can..." statement below.

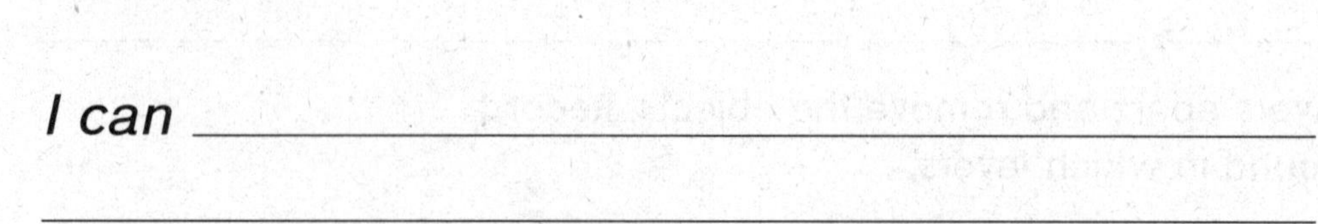

Research, Investigate, and Communicate

Exposing Rock Layers

Read *Exposing Rock Layers* on how earthquakes can change Earth's surface. Answer the questions after you have finished reading.

1. How can earthquakes expose fossils and rock layers?

2. How can other types of weathering expose fossils and rock layers?

3. How can a paleontologist determine the age of a fossil or rock layer that has been exposed by weathering or erosion?

4. How do events such as the separation of the East African Rift Valley help scientists?

Performance Task
Fossil Types

You have learned that paleontologists use clues from fossils and rock layers to learn about past environments. Like a paleontologist, you will now research different types of fossils.

Research You will use multiple sources, including your ***Science Handbook***, to research four different types of fossils: mold fossils, trace fossils, cast fossils, and true form fossils. Based on your research, you will draw one example of each and explain how your drawing illustrates that type of fossil.

Ask a Question What question will your research help to answer?

______

______

Record Data Record notes from your research on the four types of fossils in the graphic organizer below.

mold fossils	trace fossils	cast fossils	true form fossils

Analyze Data Draw an example of each type of fossil in the boxes below. Below each picture, explain what the fossil can tell us about the environment in which it lived.

mold fossil	trace fossil
cast fossil	true form fossil

Crosscutting Concepts

Patterns

1. Use the information you collected to construct an explanation of how fossils help scientists understand what the surface of Earth used to look like. Use the specific types of fossils you researched in your explanation.

Name ______________________________ Date ____________

Essential Question

What can rock formations tell us about Earth's history?

Think about the photo of the fossils you saw at the beginning of the lesson. What do you know now about what fossils can tell us about Earth's history?

____

____

____

____

____

____

____

Science and Engineering Practices

Review the "I can . . ." statement that you wrote earlier in the lesson. Explain what you have accomplished in this lesson by completing your "I did . . ." statement.

I did ______________________________

____

____

____

Patterns of Earth's Changing Features

Performance Project
Model of a Canyon

Look back at the questions that you wrote about the Grand Canyon, and the work of the park ranger, at the beginning of the module. Now that you have learned about the Earth's features, the processes that affect them, and the history of Earth's surface, you might be able to answer some of those questions. You now have the tools you need to find the answers to many more questions. It is your turn to model the processes that formed the Grand Canyon over time. You will use the data that you collect from your model as evidence to support an explanation about how the Grand Canyon formed.

Materials

- ☐ safety goggles
- ☐ scissors
- ☐ half-gallon plastic milk container
- ☐ red sand
- ☐ white sand
- ☐ yellow sand
- ☐ tan sand
- ☐ 4 books
- ☐ basin
- ☐ 2-liter container filled with water

Make a Model

BE CAREFUL Wear safety goggles during this project.

1. Cut one side off of the half-gallon milk carton and lay it on its side.
2. Fill the carton halfway with layers of red, white, yellow and tan sand.
3. Prop the side of the carton opposite the spout up with a book.
4. Place the spout end over a basin.
5. Fill a 2-liter bottle with water; begin slowly pouring water into the raised end of the carton. The water should begin flowing out of the spout. This represents a river.
6. Continue to raise the high end of the carton by adding another book. Then continue to slowly pour more water into the carton.

 Name ______________________ Date __________

7 **Record Data** Repeat step 6 two more times. How does the appearance of the "land" change?

Communicate Information

1. Use what you have observed and what you have learned to explain how the Grand Canyon formed.

Name ______________________ Date ____________

What evidence is used to learn about how Earth's surface has changed over time?

You used several different models in this module to represent landforms, weathering, erosion, fossils and rock layers. Think of a landform that you find interesting. What questions do you have about this landform? Design an investigation to answer one of your questions. Write your question and investigation below.

____

____

____

____

____

____

____

____

____

____

____

____

____

____

Explore More in Our World

Are there any questions that you still have at the end of this module? If so, use your skill of gathering evidence to find answers to your questions.

 Name ______________________ Date ____________

Natural Hazards

Science in Our World

Watch the video of people evacuating to escape the danger of a natural hazard. What questions do you have?

____

____

____

____

____

____

____

____

____

Key Vocabulary

Look and listen for these words as you learn about natural hazards.

flood	seismic wave	seismograph
seismometer	storm surge	tsunami

Name ______________________ Date ____________ MODULE OPENER

How can we avoid danger in natural hazards?

NOAH
Nurse

STEM Career Connection

Climatologist

As a climatologist, I study long-term weather patterns. I analyze large amounts of collected data to find patterns. I use these patterns to develop computer models that allow myself and other climatologists to forecast weather events. We look for long-term changes that could lead to natural disasters. Increased rainfall in an area or an increase or decrease in temperature are examples of the changes that I look for. Local officals can use the results of my research to help communities stay safe and reduce the impact of natural hazards like floods and hurricanes.

How do you think people know when a natural hazard is coming? How do they know where to go to escape its danger?

____

____

____

____

____

____

____

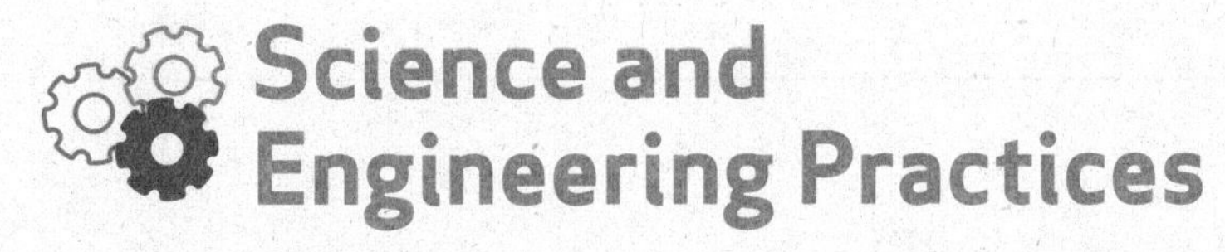

Science and Engineering Practices

I will construct explanations and design solutions to a problem.

 Name ______________________ Date __________

Earthquakes and Volcanoes

Volcanoes and Earthquakes

Two friends were talking about natural hazards such as volcanoes and earthquakes. They each had different ideas about how natural hazards affect humans. This is what they said:

Camila: *Humans cannot stop natural hazards, but they can design ways to control their impact.*

Rodney: *Humans have both the technology to prevent natural hazards and control their impact.*

Who do you agree with the most? ______________________

Explain why you agree.

____

____

____

____

Name ______________________ Date __________

Science in Our World

Watch the video of the volcano. What questions do you have?

____

____

____

____

____

____

Read the seismologist's field notes, and answer the questions on the next page.

STEM Career Connection

Seismologist

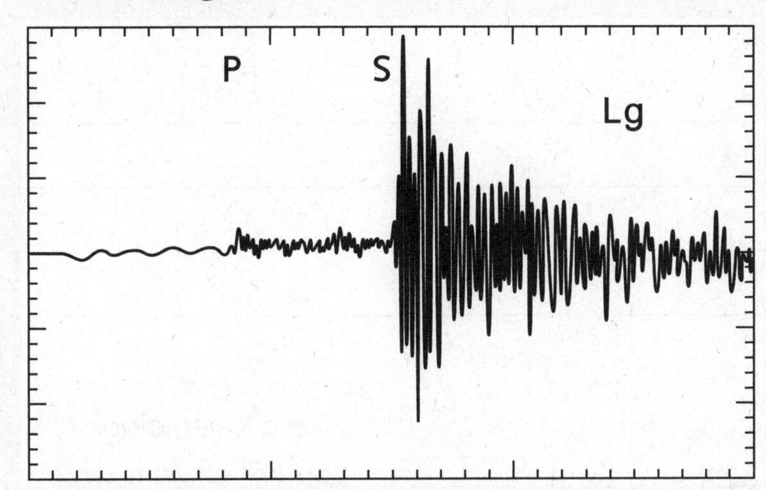

Today I was monitoring a field station near the San Andreas Fault in California for earthquake activity. The seismometer there has been recording some increased activity recently. This graph tells me about the waves produced by the recent activity.

POPPY
Park Ranger

1. What do earthquake monitoring stations do?

2. What do you think the lines on the seismograph mean?

Essential Question

How are people affected by earthquakes and volcanoes?

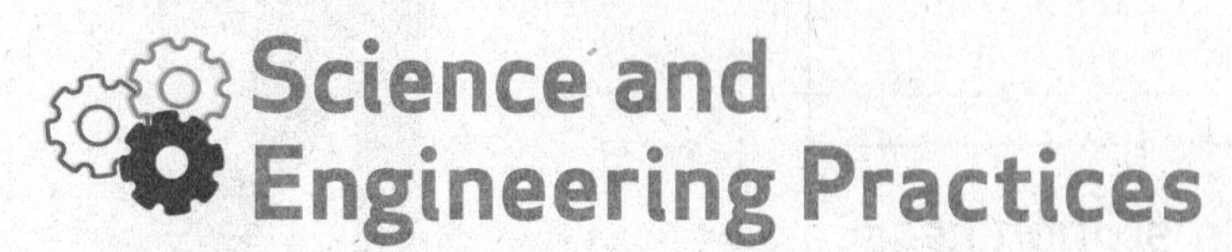

Science and Engineering Practices

I will construct explanations.

Inquiry Activity
Modeling Earthquakes

How can you model the behavior of an earthquake?

Make a Prediction What happens to rock during an earthquake? Does it behave like a solid or a liquid?

____

____

____

Materials

- ☐ safety goggles
- ☐ silicone polymer
- ☐ hammer
- ☐ gelatin dessert in 13" × 9" metal baking pan
- ☐ coffee stirrers

Carry Out an Investigation

BE CAREFUL Wear safety goggles. The chemicals in silicone polymer can be harmful if swallowed. Use all materials as directed.

1. Observe the silicone polymer. Circle the word or words that can describe the polymer:

   solid liquid wet dry

2. **Record Data** The silicone polymer represents rock found in Earth's crust. Pull the rock with a gradual, even force. Record your observations.

   ____

   ____

3. Pull the rock with a strong quick force. Record your observations.

   ____

   ____

4. Roll the silicone polymer into a ball shape. Place the silicone polymer on the floor away from your classmates and teacher. Hold the hammer firmly and strike the silicone polymer sharply. Record your observations.

   ____

   ____

   ____

 Name ______________________ Date __________

5 **Analyze Data** Which of the two forces on the polymer is most like an earthquake? Which released more energy?

____

____

____

6 **Record Data** Return the silicone polymer to your teacher, and retrieve a pan of gelatin. The gelatin represents a model of Earth's crust. Lightly tap the pan to simulate the force of an earthquake. Record your results.

____

____

____

____

7 Tap the pan of gelatin again, using more force. Record your observations.

____

____

8 To better observe the wave motion, insert equally spaced coffee stirrers into the gelatin. Repeat tapping the side of the pan with light and forceful taps. Record your observations.

____

____

____

____

____

Communicate Information

1. **Draw Conclusions** When did the silicone polymer behave like a solid? When did it behave more like a liquid? How did your results compare with your prediction?

____

____

____

____

____

____

2. Based on your observations, which type of movement do you think would cause a more destructive earthquake?

____

____

____

____

Crosscutting Concepts

Cause and Effect

3. How do you think the wave action you observed in the gelatin would affect buildings and other structures?

____

____

____

 Name ______________________ Date ____________

Obtain and Communicate Information

Vocabulary

Use these words when describing earthquakes and volcanoes.

seismic wave   seismometer   seismograph

Volcanoes and Earthquakes

Watch *Volcanoes and Earthquakes* on the processes that change Earth's surface. Answer the following questions after you have finished watching.

1. How did the San Francisco earthquake affect people?

2. What was the total economic loss due to the earthquake?

3. How did the eruption of Mount St. Helens affect people?

4. What do scientists think caused the volcano tremors?

5. How can earthquakes and volcanoes affect the environment, including plants and animals?

Volcanoes

Read *Volcanoes* on how forces within Earth can affect Earth's surface. Answer the following questions after you have finished reading.

6. Define volcano.

7. Where do volcanoes occur?

8. How do volcanoes affect plants?

Earthquakes

Read *Earthquakes* on how forces within Earth can affect Earth's surface. Answer the following questions after you have finished reading.

9. Define earthquake.

10. What causes earthquakes?

 Name ______________________ Date __________

11. What kind of damage can an earthquake do?

____

____

____

____

____

____

Science and Engineering Practices

Think about what you have learned while studying earthquakes and volcanoes. Tell how you can construct explanations by completing the "I can..." statement below.

I can ______________________________________

____

____

____

____

____

Name ______________________ Date __________ ELABORATE

Research, Investigate, and Communicate

Predicting Earthquakes and Volcanic Activity

Read *Predicting Earthquakes and Volcanic Activity* on different methods being used to predict these natural hazards.

1. What is a seismograph?

2. How does identifying and observing faults help people?

3. How is the ability to predict earthquakes and volcanic activity helpful?

Staying Safe

Read *Staying Safe* on what people can do to protect themselves from earthquakes and volcanoes.

4. What can builders do to prevent earthquake damage?

5. What have scientists learned from natural hazards like volcanoes and earthquakes?

 Name ______________________ Date __________

Performance Task

Ups and Downs of Observing a Volcano

You will make a model tiltmeter to measure movements. Tiltmeters can measure the amount of change in slope on the surface of a volcano.

Materials

- [ ] safety goggles
- [ ] pencil
- [ ] 2 foam cups
- [ ] coffee stirrer
- [ ] food coloring
- [ ] container of water
- [ ] baking pan

Make a Prediction What happens to the slope of the ground when a volcano begins to swell?

____

____

Make a Model

BE CAREFUL Wear safety goggles.

1. Using the tip of a pencil, punch a small hole into the side of a foam cup about 2.5 centimeters (cm) from the bottom of the cup. Push the coffee stirrer into the hole so it fits tightly.
2. Punch a similar hole in the second cup. Push the other end of the coffee stirrer into that hole.
3. Mix 3 drops of food coloring into a container of water. Put the connected cups into a baking pan. Then pour each cup about half full of the dyed water.
4. Carefully tilt one end of the pan. Observe what happens.
5. **Analyze Data** Continue to tilt the pan to other angles and observe the change in the level of the water. How does the tiltmeter record changes in slope?

____

____

Communicate Information

1. Explain how your prediction and the results of your model compare.

____

____

____

Crosscutting Concepts
Cause and Effect

2. What happened to the level of water in the cups when you tilted the pan?

Essential Question
How are people affected by earthquakes and volcanoes?

Think about the video of a volcano you watched at the beginning of the lesson. What do you know now about how volcanoes and earthquakes affect people?

Science and Engineering Practices

Review the "I can . . ." statement that you wrote earlier in the lesson. Explain what you have accomplished in this lesson by completing your "I did . . ." statement.

Now that you're done with the lesson, share what you did!.

I did ______________________

 Name ______________________ Date __________

Tsunamis and Floods

Preparing for Natural Hazards

Four friends were talking about the damage caused by natural hazards such as tsunamis, floods, and hurricanes. They each had different ideas about how humans can prepare for these events to lessen the damage. This is what they said:

Blake: *Humans can prepare for floods and hurricanes, but tsunamis happen too quickly to lessen the damage.*

Olivia: *Humans can prepare for hurricanes, but tsunamis and floods happen too quickly to lessen the damage.*

Pablo: *Humans can be prepared for hurricanes, floods, and tsunamis. Planning in advance can lessen the damage.*

Rita: *Hurricanes, floods, and tsunamis are very serious and harmful. Humans can't prepare for them. They have to deal with the damage after they happen.*

Who do you agree with the most? ______________________

Explain why you agree.

____

____

____

____

Science in Our World

Watch the video of a river. What questions do you have?

____

____

____

____

____

____

Read the emergency management specialist's journal entry, and answer the questions on the next page.

STEM Career Connection

Emergency Management Specialist

It is so rewarding to help people in need! Today I helped coordinate the delivery of food, water, and medicine to the victims of a hurricane in a coastal city. I also helped to set up shelters for those who had lost their homes. The victims were so grateful for help.

Not every day is as eventful as today, but we are constantly preparing for days like this. Sometimes I review emergency evacuation routes to make sure they are the most efficient. I also help communities that have been hit by a natural disaster apply for emergency funds.

POPPY
Park Ranger

Name ______________________ Date __________

1. What questions do you have about the emergency management specialist's journal entry?

____

____

____

____

2. In what ways can an emergency management specialist help people?

____

____

____

____

Essential Question

How can people prepare for floods?

____

____

____

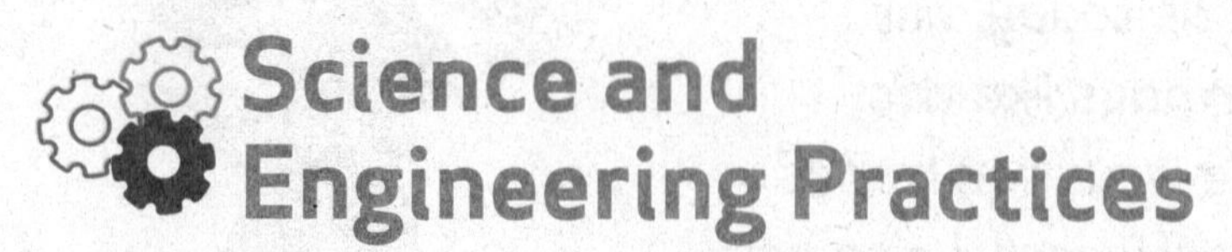

Science and Engineering Practices

I will construct explanations and design solutions.

Like an emergency management specialist, you will design solutions that help people.

Name ______________________ Date ____________ EXPLORE

Inquiry Activity
A Flooding River Model

How does a flooding river affect the surrounding area?

Make a Prediction What happens when too much water enters a river?

____

____

____

____

Materials

- ☐ safety goggles
- ☐ scissors
- ☐ aluminum baking pan
- ☐ modeling clay
- ☐ plastic tub (8 in. × 14 in. × 30 in. or 20 cm × 36 cm × 76 cm)
- ☐ book
- ☐ 2 sheets of construction paper or card stock
- ☐ paper cup
- ☐ water
- ☐ paper towels

Make a Model

BE CAREFUL Wear safety goggles and use materials as directed.

1. Use the point of the scissors to poke drain holes in one end of the baking pan, or cut off the entire end of the pan.
2. Pack the pan with clay about 1 centimeter (cm) thick.
3. Place the large plastic tub on a table. Prop one end of the tub up with a book. Place the clay-filled pan in the tub at the higher end.
4. Use your finger to carve a fairly straight, narrow river channel down the length of the pan.
5. Construct ten 2-cm houses out of construction paper and place them in the model. Put some near the river, and some away from the river.
6. **Record Data** Pour ½ cup of water slowly into the channel at the higher end. Record your observations, including the number of damaged houses.

____

____

____

7. Dry your model with paper towels. Replace any damaged houses with new ones in approximately the same places as before.

 Name ____________________ Date __________

8 **Record Data** Pour ½ cup of water quickly into the model, and record your observations below.

9 Dry your model with paper towels. Replace any damaged houses with new ones in approximately the same places as before. Add an obstruction made from clay that blocks the water flow about halfway down the river channel.

10 Repeat steps 6 and 7.

11 Repeat step 8.

Communicate Information

1. How did your prediction compare with the results of this investigation?

Name ______________________ Date __________ EXPLAIN

Obtain and Communicate Information

Vocabulary

Use these words when describing natural hazards.

tsunami flood storm surge

Hurricanes, Tsunamis, and Floods

Watch *Hurricanes, Tsunamis, and Floods* on how these natural disasters affect people. Answer the following questions after you have finished watching.

1. What part of hurricanes causes damage?

____

____

2. What causes damage in a tsunami?

____

3. What can cause flooding?

____

4. What do these three types of natural hazards have in common?

____

5. Which natural hazard would be easiest to prepare for? Explain your choice using what you have learned so far.

____

____

____

Name ______________________ Date __________

Understanding and Tracking Natural Hazards

Read *Understanding and Tracking Natural Hazards.* Answer the following questions after you have finished reading.

6. Where do tropical storms and hurricanes form?

7. When is a tropical storm considered a hurricane?

8. Explain the hurricane classification system.

9. How do scientists track hurricanes?

10. Why are floods dangerous?

11. In what situation can a flood be helpful?

Tsunami!

Read *Tsunami!* Answer the following questions after you have finished reading.

12. What events can cause a tsunami?

____

____

____

13. What happens when a tsunami hits the shore?

____

____

____

____

14. What should people do if there is a tsunami warning?

____

____

15. What areas are in danger from tsunamis?

____

____

____

____

16. How do scientists know when a tsunami is coming?

____

____

____

Name ______________________ Date ____________

FOLDABLES®

Cut out the Notebook Foldables tabs given to you by your teacher. Glue the anchor tabs as shown below. Use what you have learned to explain how natural disasters might threaten the house.

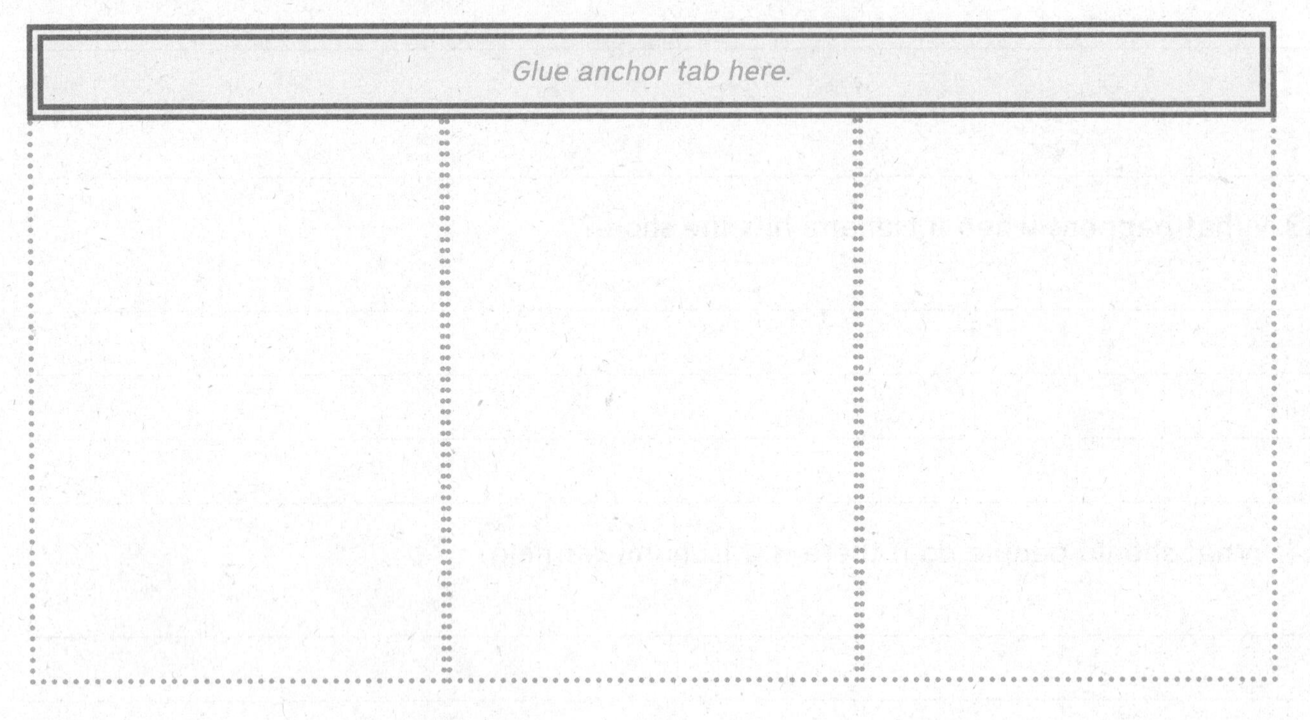

Name ______________________ Date __________ EXPLAIN

Historical Tsunamis Around the World

Explore the Digital Interactive *Historical Tsunamis Around the World.* Answer the questions after you have finished.

17. Describe the cause, location, and wave height of the March 1964 tsunami.

____

18. Describe the cause, location, and wave height of the December 2004 tsunami.

____

Crosscutting Concepts
Cause and Effect

19. Where are future tsunamis likely to happen? Support your answer with evidence.

____

____

____

____

Science and Engineering Practices

Use examples from the lesson to explain what you can do!

Think about what you have learned about natural hazards. Express how you can construct explanations and design solutions by completing the "I can..." statement below.

I can ____

____

____

 Name ______________________ Date __________

Research, Investigate, and Communicate

Preparing for Natural Hazards

Read *Preparing for Natural Hazards.* Answer the following questions after you have finished reading.

1. How can monitoring and an early warning system help people prepare for natural hazards?

____

____

____

____

2. How can buildings be designed to withstand the effects of natural hazards?

____

____

____

____

3. What natural hazard is the most likely threat for each city? Write the name of the natural hazard beside the name of the city. If a city has multiple hazards that are likely, then you may list more than one. Use what you have learned about each natural hazard. You will want to use a map to locate each city to help you determine the threat of each natural hazard.

Minneapolis, MN		**Seattle, WA**	
Phoenix, AZ		**Tulsa, OK**	
Savannah, GA			

Name ______________________ Date ____________ EVALUATE

Performance Task
Flooding River: A Solution

You will revisit your river model from the beginning of the lesson. Your goal will be to design a solution to reduce the effects of flooding.

Materials
- ☐ river model from beginning of lesson
- ☐ various materials available in the classroom

Make a Prediction How could the negative effects of a flooding river be reduced?

____

____

____

Design a Solution

1. Review your observations from the A Flooding River Model activity. Brainstorm some ideas for reducing the effects that flooding had on the model homes. You might consider making changes to the river bed or to the model homes. List your ideas below:

____

____

____

2. Design, draw, and explain your solution.

3 Record Data Make the changes to your model to reflect your design. Then test your design. Record your observations below.

Communicate Information

1. Do the results of your tests support your prediction?

2. Based on your test, what changes would you make to improve your design?

3. Do you think making changes to the river or making changes to the structures near the river would better reduce the effects of flooding? Explain your answer using what you have learned in this lesson.

____

____

Essential Question

How can people prepare for floods?

Think about the video of a flooding river at the beginning of the lesson. What have you learned about ways that people can prepare for floods?

____

____

____

____

____

____

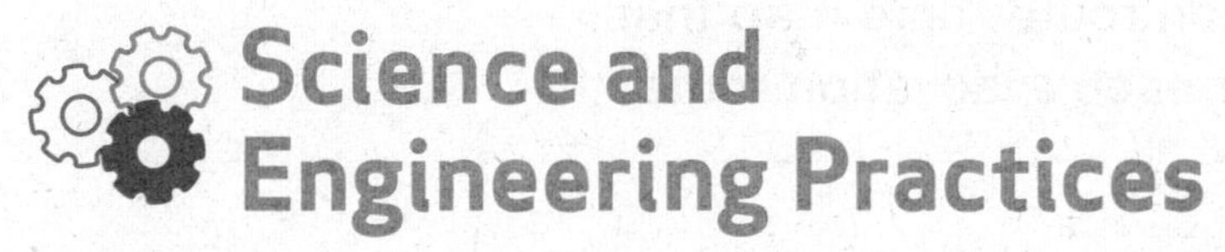

Science and Engineering Practices

Now that you're done with the lesson, share what you did!

Review the "I can . . ." statement you wrote earlier in the lesson. Explain what you have accomplished in this lesson by completing the "I did . . ." statement.

I did ____

____

____

____

 Name ______________________ Date ____________

Natural Hazards

Performance Project

Natural Disaster Safety

Review the questions you wrote about people evacuating at the beginning of the module.

You learned about different types of natural hazards in this module. For some hazards, such as flooding and volcanoes, evacuating your town may be necessary for survival. For other natural hazards, taking shelter indoors is the best way to stay safe. This would include such hazards as tornados, electrical storms, severe winter storms, and extreme heat.

Select and label a natural hazard that could affect your area and would require evacuation for safety. Sometimes routes get blocked, so it's important to have a back-up plan. You are going to plan and label three possible evacuation routes on a map that you draw below. Use a different color for each evacuation route.

Name ______________________ Date ____________

How can we avoid danger in natural disasters?

Design a building that can withstand the impact of the natural hazard that you have chosen. Draw and label your design below. Include a short explanation of the type of natural disaster your building is designed to withstand.

____

____

____

Explore More in Our World

Did you learn the answers to all of your questions from the beginning of the module? If not, how could you design an experiment or conduct research to help answer them?

 Name ____________________ Date __________

Energy from Natural Resources

Science in Our World

Look at the photo of the biofuel-powered vehicle. What questions do you have?

____

____

____

____

____

____

____

____

abc Key Vocabulary

Look and listen for these words as you learn about energy from natural resources.

alternative energy source	biomass	biomass conversion
conservation	fossil fuel	geothermal
hydroelectricity	natural resource	nonrenewable resource
pollution	renewable resource	replenish

Name ______________________ Date ____________

How can we choose energy resources in a wise way?

CJ
Statistician

STEM Career Connection

Biochemist

As a biochemist, I study the chemistry of living things. I work in the biofuels industry, so my research is focused on finding the best way to turn plants or other living things into fuel. I am currently conducting experiments on algae. My experiments test how effectively different technologies can break down algae to turn it into fuel for vehicles. This is an exciting field because new technology is developed all the time as people look for alternatives to fossil fuels. Algae is an energy source that can replace itself in days as opposed to the millions of years it takes for fossil fuels to renew. My work could change the way the world thinks about fuel.

What factors should you consider when choosing a resource?

____

____

____

____

____

____

Science and Engineering Practices

I will obtain, evaluate, and communicate information.

 Name ______________________ Date __________

Energy from Nonrenewable Resources

Energy from Natural Resources

Four friends were talking about nonrenewable energy resources such as natural gas, coal, and oil. They each had different ideas about these energy resources. This is what they said:

Jake: *Coal and oil can harm the environment. Natural gas is natural, so it doesn't harm the environment.*

Kim: *Oil can harm the environment because it can get into water. Coal and natural gas don't harm the environment.*

Alisha: *Coal, oil, and natural gas are all safe for the environment because they are natural resources.*

Lucas: *Coal, oil, and natural gas can all harm the environment.*

Who do you agree with most? ______________________

Explain why you agree.

____

____

____

____

Name ________________ Date ________ ENGAGE

Science in Our World

Look at the photo of a coal mine. What questions do you have?

____

____

____

____

____

____

Read the nuclear engineer's journal entry, and answer the questions on the next page.

STEM Career Connection

Nuclear Engineer

I work at a nuclear power plant that delivers electricity to many homes and businesses. It is my job to make sure that the nuclear reactor continues operating properly. We use a certain kind of uranium as fuel. Uranium is a nonrenewable resource. Energy is stored in the nucleus, or center, of the uranium atoms. In the nuclear reactor, the uranium atoms are split apart to release stored energy. My job includes monitoring the supply of uranium and using computer models to predict how changes will affect the nuclear reactor.

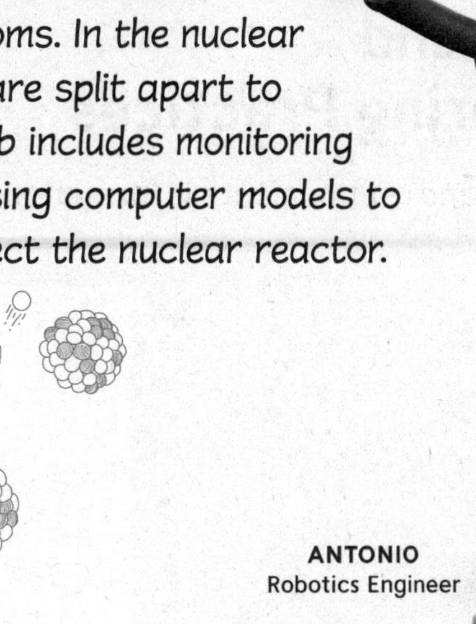

Like coal, uranium is a natural resource. Nuclear engineers help supply homes and businesses with electricity.

ANTONIO
Robotics Engineer

 Name ______________________ Date __________

1. How would you explain what is happening in the image?

____

____

____

2. How is the energy stored in uranium atoms used?

____

____

____

Essential Question

What are nonrenewable resources?

____

____

____

____

____

Science and Engineering Practices

I will obtain, evaluate, and communicate information.

Like a nuclear engineer, you will obtain information about nonrenewable resources.

Inquiry Activity
Limited Resources

Materials

- ☐ 60 pennies, beads, or other small objects

How plentiful are the coal, oil, and natural gas reserves that we depend on for energy? You will investigate what happens when we use resources that are not replaced as quickly as they are used.

Make a Prediction

What factors control how long our energy reserves will last? Which factor can be controlled?

____

____

____

____

____

Carry Out an Investigation

1. Work in groups of 4. Sit with your group.
2. Place 20 pennies in the center of the group.
3. In each round, each student takes at least one penny from the center. One penny is the basic energy need. Students may take as many as they want; however, the goal is to not run out during a round.
4. Between rounds, the teacher will add to each group half the number of pennies that were taken. For example, if a group took 10 pennies, the teacher will add 5. The teacher may round up or down if an odd number is left.
5. Groups will complete 6 rounds, unless they run out of resources before then.

 Name ______________________ Date __________

Communicate Information

1. What happened to the supply of pennies over time?

2. How do the pennies represent nonrenewable resources?

3. How did the different groups manage their resources?

4. **Analyze Data** What would eventually happen to the pennies if you continued playing?

Crosscutting Concepts

Cause and Effect

5. What will eventually happen to our nonrenewable natural resources?

6. What factor controls how soon they will run out? Explain.

Name ______________________ Date ____________ EXPLAIN

Obtain and Communicate Information

Vocabulary

Use these words when describing Earth's nonrenewable resources.

conservation	fossil fuel	natural resource
nonrenewable resource	pollution	replenish

Using Nonrenewable Resources

Watch *Using Nonrenewable Resources* on Earth's natural resources. Answer the following questions after you have finished watching.

1. What are the two types of natural resources?

____

____

2. Give an example of a renewable resource. What makes it renewable?

____

____

____

3. Give an example of a nonrenewable resource. What makes it nonrenewable?

____

____

____

4. What are some other nonrenewable resources?

____

____

EXPLAIN Name ____________________ Date __________

Fossil Fuels

Read pages 126–127 in the ***Science Handbook.*** Answer the following questions after you have finished reading.

5. What is a fossil fuel?

6. How did fossil fuels form?

7. How are fossil fuels used?

Effects of Using Fossil Fuels

Read page 130 in the ***Science Handbook.*** Answer the following questions after you have finished reading.

8. What are the causes of air pollution?

9. How can water and land become polluted?

Name ____________________ Date __________ EXPLAIN

Inquiry Activity
Oil Spill Clean-Up

You will model an oil spill and try different methods for cleaning it up.

Make a Prediction Oil spills harm ecosystems. How will you be able to clean up an oil spill?

____

____

Materials

- ☐ aluminum pie pan or other container
- ☐ water
- ☐ rocks, toy plants and toy animals
- ☐ dark olive oil
- ☐ bird feather from hobby shop
- ☐ paper towels
- ☐ grease-fighting dish detergent
- ☐ plastic spoon
- ☐ cotton balls
- ☐ pieces of sponge
- ☐ chenille stems
- ☐ straws

Carry Out an Investigation

1. Pour 1/4 to 1/2 inches of water in the pie pan. Arrange the rocks and toy plants and animals to model an ecosystem.
2. Your teacher will add an "oil spill" (about 2 tablespoons of dark olive oil) to your ecosystem.
3. Dip the bird feather into the oily water.
4. Try wiping the oil off of the feather with a paper towel. Try rinsing it off in the sink.
5. Use a drop of detergent and some water to wash the feather, and then gently dry it with a paper towel.
6. Go back to the pie pan. Use the spoon to try to skim the oil off of the water.
7. **Record Data** Record your observations. What happened when oil was added to your model ecosystem? How did it change the bird feather? How did the different methods work to clean it up?

____

____

____

____

____

____

Name ______________________ Date ________

8 Using the other materials available, come up with a plan to clean up your oil spill. Outline your plan here, and share it with your teacher before you begin.

Communicate Information

10. How well did your plan work? Explain.

11. How did your prediction compare with the results of the activity?

Conservation

Read pages 132–133 in the ***Science Handbook.*** Answer the following questions after you have finished reading.

12. What are some ways people can conserve resources?

13. Give an example of how to reduce resource use.

14. Define *recycle*. Give an example.

FOLDABLES®

Cut out the Notebook Foldables tabs given to you by your teacher. Glue the anchor tabs as shown below. Use what you have learned to define and apply each vocabulary word to the picture of the oil pumpjack.

Glue anchor tab here

 Name ______________________ Date __________

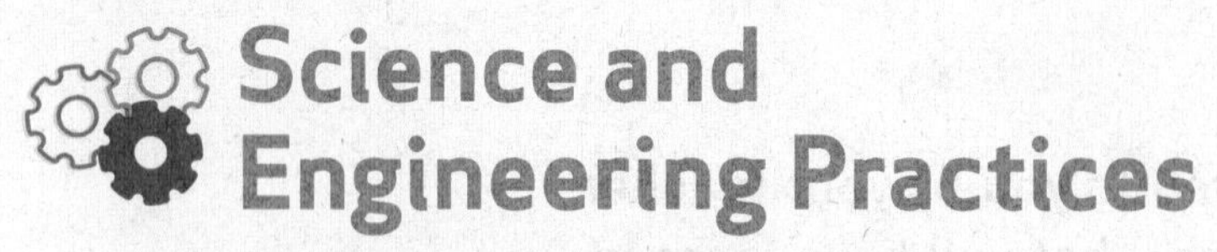

Science and Engineering Practices

Think about the information you have obtained from different sources about the effects of using nonrenewable resources on the environment. Complete the "I can..." statement below.

I can ____

____

____

____

____

____

Research, Investigate, and Communicate

Converting Units

When people shop for a car, they often consider how much gas it uses. A fuel-efficient car travels a longer distance on a smaller amount of gasoline. This saves the driver money and also helps conserve oil. In the United States, we measure the amount of gasoline that a car uses in miles per gallon (mpg). Scientists and people in most other countries use the units kilometers per liter (kmpl).

1. To convert kmpl to mpg, multiply the number of kmpl by 2.352. For example, 12 kmpl × 2.352 = 28.224 mpg
2. To convert mpg to kmpl, multiply the number of mpg by 0.425. For example, 40 mpg × 0.425 = 17.000 kmpl

Communicate Information

1. Sam's car gets 25 mpg. Jasmine's car gets 29 mpg. Which car is more fuel-efficient?

2. Lori's car gets 36 mpg. Henry's car gets 9 kmpl. How much gas does Henry's car use in miles per gallon? Which car is more fuel-efficient?

3. Maria drove 64 kilometers on 4 liters of gas. Jerry drove 58 miles on 2 gallons of gas. How many kilometers per liter of gas can each car drive? Which car is more fuel-efficient?

Performance Task

Energy Usage Investigation

People use energy in the form of electricity every day. You may play video games, watch TV, or use a computer. Electricity reaches your house through power lines. Power lines connect your home to a power plant that generates electricity. The majority of this energy is converted from nonrenewable resources. Your family pays for the amount of electricity that is used. The electric company reads a meter to see how much electricity has been used, and then sends a bill. Your family may also use energy in the form of gas, another nonrenewable resource. The vehicle that your family uses to drive around may get its energy by burning gasoline. Some homes are heated with natural gas.

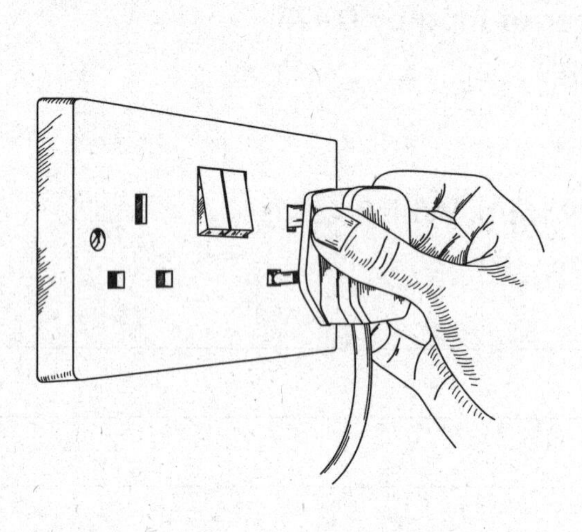

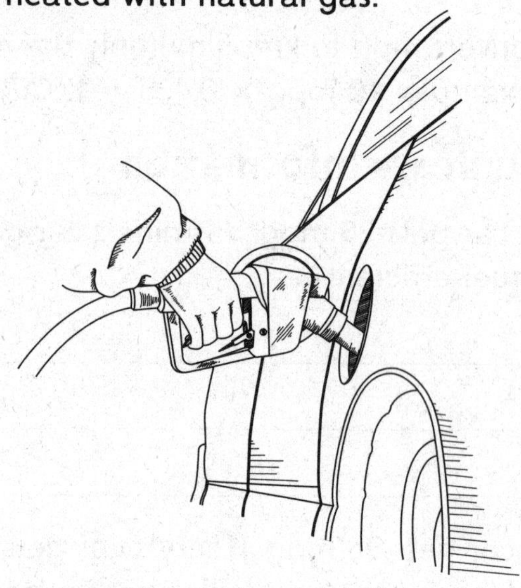

Carry Out an Investigation

1. Gather information about energy use at your home.

   How many lightbulbs do you have in your home that you use for several hours each day?

   ______

   How many people live in your home?

   ______

2. Use the Internet to help you research the population of your town and the energy sources used by your electric company.

   What source or sources of energy are used by your electric company?

   ______

What is the population of your town?

____

3 Calculate how many energy units the lightbulbs in your home use and how much the energy costs. Assume that each lightbulb from step 1 uses 1 kilowatt-hour (kwh) of energy per day if it is on for 10 hours and 1 kwh costs $0.10. Use the space below to show your work.

4 **Analyze Data** Use a separate sheet of graph paper to create a bar graph comparing your classmates' daily energy usage at home. Be sure to label the X and Y axes and the data you input. Include your own data and data from at least 4 other students.

5 Assume that each person in your town uses the same amount of energy as your family members. Calculate how much energy is used by your town in one day. Use the space below to show your work.

Communicate Information

1. How does your family's energy usage compare to your classmates'?

____

____

2. Think about the number of hours your family actually leaves the lights on in your home. Outline a plan to help reduce weekly energy usage. Include how your family members will be involved.

____

____

Glue your graph here.

 Name ______________________ Date __________

3. Using your family's daily energy use, calculate how much energy each person in your home uses each day. Divide the total number of energy units by the number of people living in your home. This will be your average number of energy units per person. Use the space below to show your work.

Crosscutting Concepts
Cause and Effect

4. Why is it important to reduce energy usage?

Name ______________________ Date ____________ EVALUATE

Essential Question

What are nonrenewable resources?

Think about the photo of a coal mine at the beginning of the lesson. What do you know now about nonrenewable resources and how people use them?

____

____

____

____

Science and Engineering Practices

Review the "I can . . ." statement you wrote earlier in the lesson. Explain what you have accomplished in this lesson by completing the "I did . . ." statement.

I did ____

____

____

____

____

____

____

____

____

____

____

____

____

 Name ____________________ Date __________

Energy from Renewable Resources

Renewable Resources

Some energy resources can be used over again without being depleted, or used up. These are called renewable energy resources. Draw an X over any box that contains a renewable energy resource.

Oil	**Wood**	**Wind**
Sun	**Water**	**Coal**
Natural Gas	**Living Things**	**Heat from Inside Earth**
Corn	**Fossil Fuels**	**Gasoline**

Explain your thinking. How did you decide if something is a renewable energy resource?

____

____

____

____

____

____

____

Name ______________________ Date ______________

Science in Our World

Watch the video of a waterfall. What questions do you have?

____

____

____

____

____

____

Read the hydroelectric plant technician's field notes, and answer the question on the next page.

STEM Career Connection

Hydroelectric Plant Technician

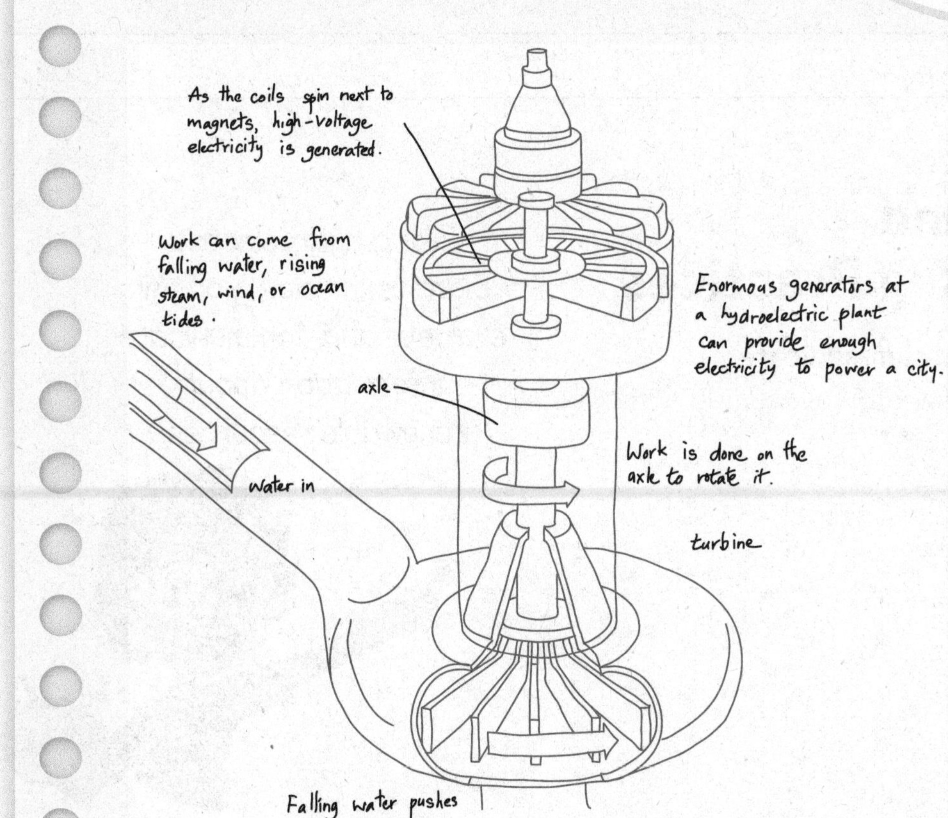

ANTONIO
Robotics Engineer

 Name ______________________ Date ________

1. How would you explain the way water can be used to make electricity? Refer to the hydroelectric plant technician's notes.

Essential Question

How are renewable resources used as energy?

Science and Engineering Practices

I will obtain, evaluate, and communicate information.

Like a hydroelectric plant technician, you will evaluate and communicate information about renewable resources.

Inquiry Activity
Renewable Resources

Materials

- [ ] 100 pennies, beads, or other small objects

Can renewable resources run out?

Make a Prediction You will model how energy resources like hydroelectric, solar, and wind energy can be replenished over time. How can you make energy resources last the longest?

____

____

____

____

____

Carry Out an Investigation

1. Work in groups of 4. Sit with your group.
2. Place 20 pennies in the center of the group.
3. In each round, each student takes at least one penny from the center. One penny is the basic energy need. Students may take as many as they want; however, the goal is to not run out of pennies during a round.
4. Between rounds, the teacher will add to each group's resources. Groups will get the same number of pennies that are left in the center pile. For example, if a group has 10 pennies left, the teacher will add 10.
5. Groups will complete 6 rounds unless they run out of resources before then.

Communicate Information

1. How do the pennies represent renewable resources?

2. How did the different groups manage their resources?

3. How did the outcome of this activity differ from the Limited Resources activity in Lesson 1?

4. What would eventually happen to the pennies if you continued playing?

Crosscutting Concepts

Cause and Effect

5. Will renewable resources ever run out?

6. How can using renewable resources make nonrenewable resources last longer?

Name ______________________ Date ____________ EXPLAIN

Obtain and Communicate Information

Vocabulary

Use these words when describing renewable resources.

alternative energy source	biomass	biomass conversion
geothermal	hydroelectricity	solar power
renewable resource		

Renewable and Alternative Energy Sources

Read pages 128–129 in the ***Science Handbook***. Answer the following questions after you have finished reading.

1. What is a renewable resource?

____

____

2. What are some examples of renewable resources?

____

____

3. Where do plants get their energy?

____

4. What are alternative energy sources?

____

____

 Name ______________________ Date __________

Solar Energy

Read *Solar Energy* about a renewable energy resource. Answer the following questions after you have finished reading.

5. What are some ways we use solar energy?

____

____

6. What is insolation?

____

____

7. In what ways can solar energy be used to generate electricity?

____

____

____

____

8. When is solar energy not available?

____

Renewable Hydroelectric and Geothermal Energy

Explore the Digital Interactive *Renewable Hydroelectric and Geothermal Energy.* Answer the following questions after you have finished.

9. What is the purpose of the turbine in the hydroelectric power plant?

____

10. What does the generator do in the hydroelectric power plant?

____

11. How is energy carried to homes and businesses?

____

12. What happens to the water after it goes through the plant?

____

Biomass: An Alternative Energy

Read *Biomass: An Alternative Energy* about how living things can be used as a renewable energy source. Answer the following questions after you have finished reading.

13. What is biomass?

14. What is biofuel?

15. What is bioethanol made from?

16. What is biogas made from?

17. What is a benefit of using algae instead of corn as a biofuel?

Name ______________________ Date __________

FOLDABLES®

Cut out the Notebook Foldables tabs given to you by your teacher. Glue the anchor tabs as shown below. Use what you have learned to describe each type of energy that can come from the renewable resources shown in the picture.

Glue anchor tab here

Science and Engineering Practices

Think about what you have learned about renewable resources. Express how you can obtain, evaluate, and communicate information about renewable energy resources by completing the "I can..." statement below.

I can ____

____

____

____

____

____

____

____

 Name ______________________ Date __________

Research, Investigate, and Communicate

Pittsburgh's Transformation

Read *Pittsburgh's Transformation* about a city that cleaned up pollution over time. Answer the following questions after you have finished reading.

Crosscutting Concepts

Cause and Effect

1. Why was Pittsburgh's air so polluted?

2. Why was the city's river polluted?

3. What did people think about pollution during the 1800s?

4. What changed the air quality in Pittsburgh?

5. How has Pittsburgh changed? Give examples.

Energy Decision

You and a group of three friends in your neighborhood decide to go to a movie. The movie theater is too far away to walk or ride your bike. You must choose between all riding separately to the movie with your parents, carpooling so that only one parent has to drive, or taking a city bus with your friends and a parent. You must make a decision about the best use of energy to get you and your friends to the movie theater.

6. Write a paragraph explaining your decision. Use evidence from the module to support your choice.

 Name ____________________ Date __________

Energy Supply

Investigate where our energy supply comes from by conducting the simulation. Answer the following questions after you have finished.

Communicate Information

7. Explain how you managed the city's power supply.

8. What was the biggest problem with using coal power plants to generate electricity?

9. What was the biggest problem with using wind turbines to generate electricity?

Name ______________________ Date __________ EVALUATE

Performance Task

Renewable Energy Campaign

You just learned about different energy sources and the benefits of using renewable resources. You can now make educated decisions about many of the resources that you use. In this activity, you will create a campaign to teach others about a renewable energy source. Use information from the lesson Science Files, leveled readers, and activities to develop a presentation that explains why people should use a particular type of renewable energy. Conduct research to add to your knowledge and gather evidence to support your claim.

Communicate Information

1. Which renewable energy source have you chosen? Why did you choose this source? If your decision changes during your research, edit this section to reflect your new choice.

2. Research your chosen renewable energy source. Use the space below to record information that you will use for your presentation.

3. Make an outline for your presentation. Be sure to include the evidence that you will use to support your claim.

4. What further information would you like to know about your energy source? Include at least two questions.

5. Was your presentation successful in convincing your audience to use your chosen renewable energy source? Explain.

Essential Question

How are renewable resources used as energy?

Think about the waterfall video at the beginning of the lesson. What do you know now about how renewable resources are used for energy?

____

____

____

____

____

____

Science and Engineering Practices

Review the "I can . . ." statement you wrote earlier in the lesson. Explain what you have accomplished in this lesson by completing the "I did . . ." statement.

I did ______________________________

____

____

____

____

____

____

____

____

____

____

 Name ______________________ Date __________

Energy from Natural Resources

Performance Project

Making Wise Choices

Look back at the module opener about the biofuel-powered bus. Think about the factors that are important to consider when choosing an energy resource. Biofuel, as you know, is a renewable energy resource. You also know that gasoline is a nonrenewable energy resource. People may wonder if biofuel a good alternative to gasoline as an energy resource. Why is it important to make wise energy resource choices? Explain your answer using information that you have gathered in this module.

Name ______________________ Date ____________

How can we choose energy resources in a wise way?

In this activity, you will design a process for rating different resources so that people can make wise decisions. You will use research materials and what you have learned throughout the module to develop criteria that will help people decide among resources. Your plan should include at least five different criteria, or factors to consider, and a ranking system.

Explore More in Our World

Did you learn the answers to all of your questions from the beginning of the module? If not, how could you design an experiment or conduct research to help answer them?

An Interview with

Dinah Zike Explaining Visual Kinesthetic Vocabulary®, or VKVs®

What are VKVs and who needs them?

“VKVs are flashcards that animate words by kinesthetically focusing on their structure, use, and meaning. VKVs are beneficial not only to students learning the specialized vocabulary of a content area, but also to students learning the vocabulary of a second language.”

Dinah Zike | Educational Consultant
Dinah-Might Activities, Inc. – San Antonio, Texas

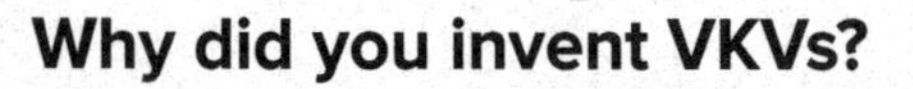

Why did you invent VKVs?

“Twenty years ago, I began designing flashcards that would accomplish the same thing with academic vocabulary and cognates that Foldables® do with general information, concepts, and ideas—make them a visual, kinesthetic, and memorable experience.”

Dinah Zike's Visual Kinesthetic Vocabulary®

I had three goals in mind:

- **Making two-dimensional flashcards three-dimensional**
- **Designing flashcards that allow one or more parts of a word or phrase to be manipulated and changed to form numerous terms based upon a commonality**
- **Using one sheet or strip of paper to make purposefully shaped flashcards that were neither glued nor stapled, but could be folded to the same height, making them easy to stack and store**

Why are VKVs important in today's classroom?

“At the beginning of this century, research and reports indicated the importance of vocabulary to overall academic achievement. This research resulted in a more comprehensive teaching of academic vocabulary and a focus on the use of cognates to help students learn a second language. Teachers know the importance of using a variety of strategies to teach vocabulary to a diverse population of students. VKVs function as one of those strategies.”

An Interview with

Dinah Zike Explaining Visual Kinesthetic Vocabulary®, or VKVs®

How are VKVs used to teach content vocabulary to EL students?

“VKVs can be used to show the similarities between cognates in Spanish and English. For example, by folding and unfolding specially designed VKVs, students can experience English terms in one color and Spanish in a second color on the same flashcard while noting the similarities in their roots.”

How are VKVs used to teach content vocabulary?

“As an example, let's look at content terms based upon the combining form *–vore*. Within a unit of study, students might use a VKV to kinesthetically and visually interact with the terms *herbivore, carnivore,* and *omnivore*. Students note that *–vore* is common to all three words and it means “one that eats” meat, plants, or both depending on the root word that precedes it on the VKV. When the term *insectivore* is introduced in a classroom discussion, students have a foundation for understanding the term based upon their VKV experiences. And hopefully, if students encounter the term *frugivore* at some point in their future, they will still relate the *–vore* to diet, and possibly use the context of the word's use to determine it relates to a diet of fruit.”

Dinah Zike's book, *Foldables, Notebook Foldables, & VKVs for Spelling and Vocabulary 4th-12th,* won a Teachers' Choice Award in 2011 for “instructional value, ease of use, quality, and innovation”; it has become a popular methods resource for teaching and learning vocabulary.

cut on all dashed lines

fold on all solid lines

potential energy

_______ is the energy an object has because it is moving.

_______ is the energy that is stored inside an object.

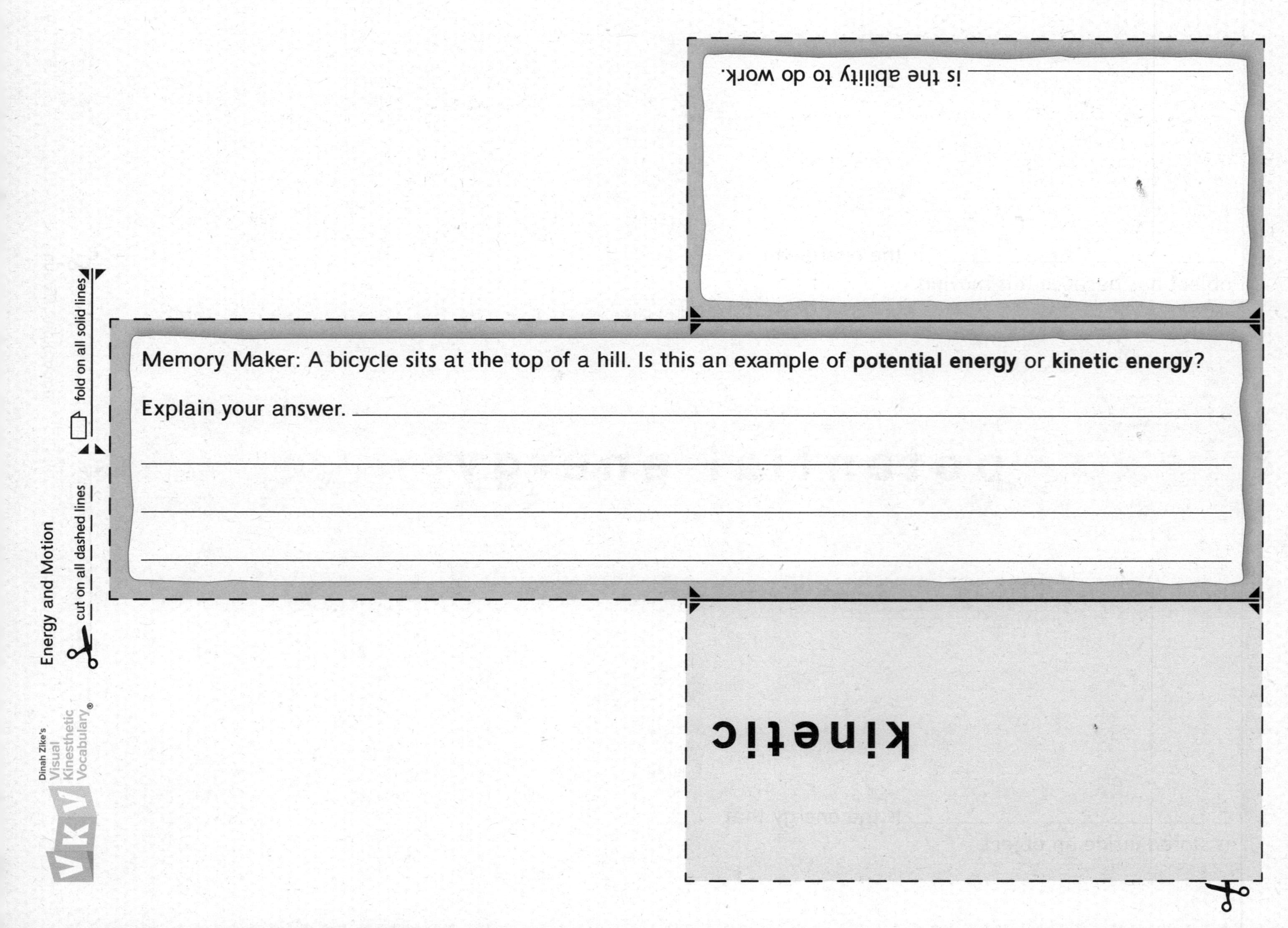

Memory Maker: A bicycle sits at the top of a hill. Is this an example of potential energy or kinetic energy?
Explain your answer.
______ is the ability to do work.
kinetic
Dinah Zike's
VKV
Visual Kinesthetic Vocabulary®
Energy and Motion
cut on all dashed lines
fold on all solid lines

cut on all dashed lines

fold on all solid lines

unbalanced force

______ are forces that act together on an object without changing its motion.

______ are forces that do not cancel each other out and that cause an object to change its motion.

cut on all dashed lines
fold on all solid lines

s

Memory Maker: Use your own words to explain the difference between **balanced forces** and **unbalanced forces**. ____________________

____

____

____

balanced

____________ is a push or pull.

Dinah Zike's Visual Kinesthetic Vocabulary®

Transfer of Energy

cut on all dashed lines

fold on all solid lines

circuit

A ________ is a flow of electricity through a conductor.

A ________ is a material through which heat or electricity flows easily.

________ is the transfer of thermal energy between two objects that are touching.

conduction

Memory Maker: The word part **circ** in **circuit** means "circle." Look at the diagram on the other side of this card. What creates the circle in a **circuit**? ______

electric

or

Memory Maker: A **conductor** is "one who conducts." **Conduction** is "the act of conducting." What does **conduct** mean? (Note: **Conduct** is a multiple-meaning word. Write its science meaning on the following lines.)

cut on all dashed lines

fold on all solid lines

Transfer of Energy

cut on all dashed lines

fold on all solid lines

________________ is the movement of heat from a warmer object to a cooler object.

conservation of energy

________________ is a principle in physics that states that energy can neither be created nor destroyed and that the total energy of a system by itself remains constant.

Dinah Zike's Visual Kinesthetic Vocabulary®

Transfer of Energy

cut on all dashed lines

fold on all solid lines

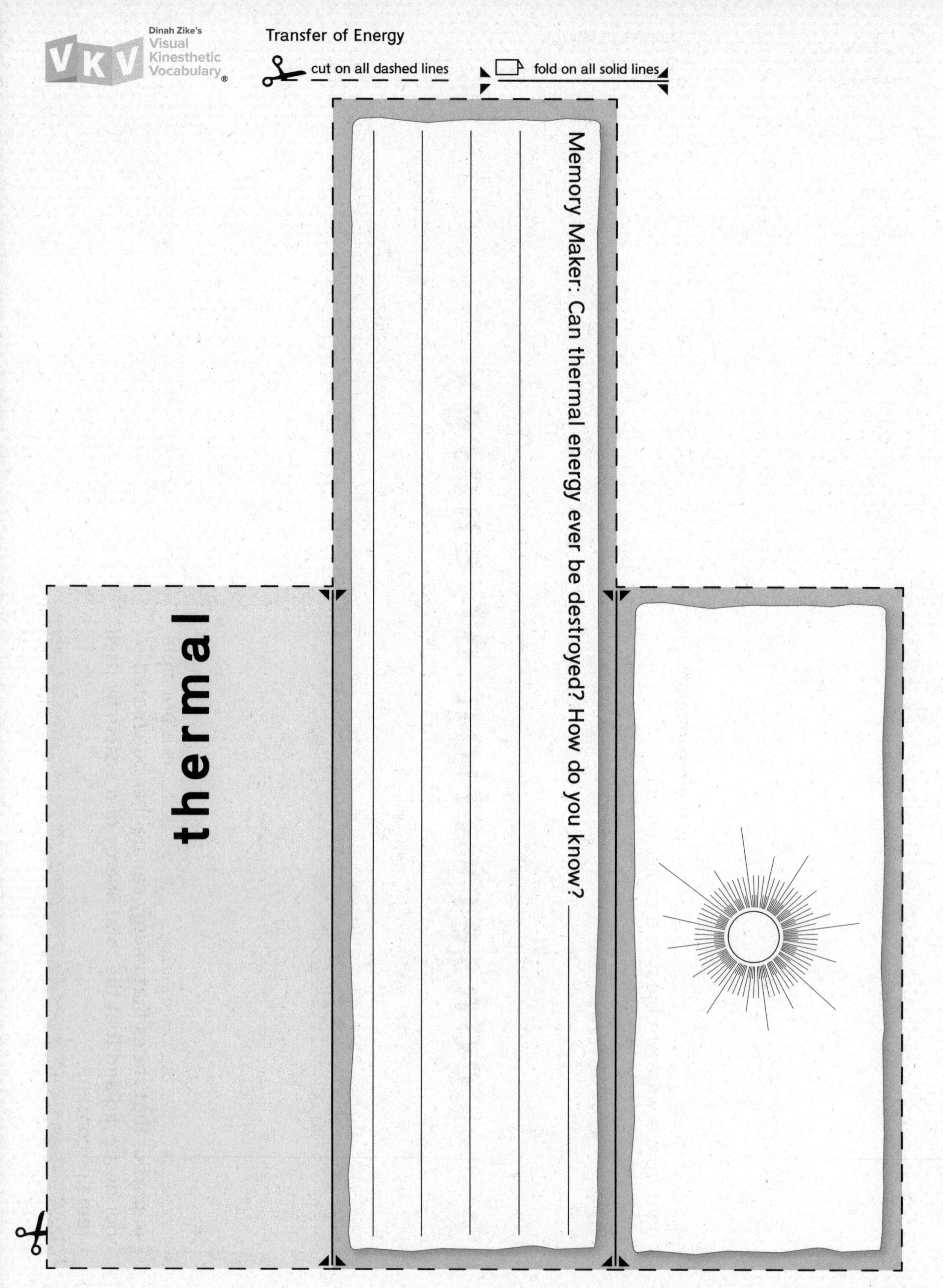

cut on all dashed lines

fold on all solid lines

tropism

_________________ is a movement or growing in a particular direction that is made by a living thing in response to light.

An _________________ is an internal supporting structure.

An _________________ is a hard covering that protects the bodies of some invertebrates.

exoskeleton

cut on all dashed lines

fold on all solid lines

Memory Maker: The word **phototropism** has two parts: **photo** and **tropism**. If **photo** means "light," what does **tropism** mean? ______________

____

____

____

photo

endo

Memory Maker: The word part **endo** means "inside" and the word part **exo** means "outside." How does knowing the meanings of these word parts help you remember the meanings of **endoskeleton** and **exoskeleton**?

____

____

____

cut on all dashed lines

fold on all solid lines

vertebrate

A ______ is an animal with a backbone.

An ______ is an animal without a backbone.

______ is the using and releasing of energy in a cell.

A ______ is the organ system that brings oxygen to body cells and removes waste gas.

respiratory system

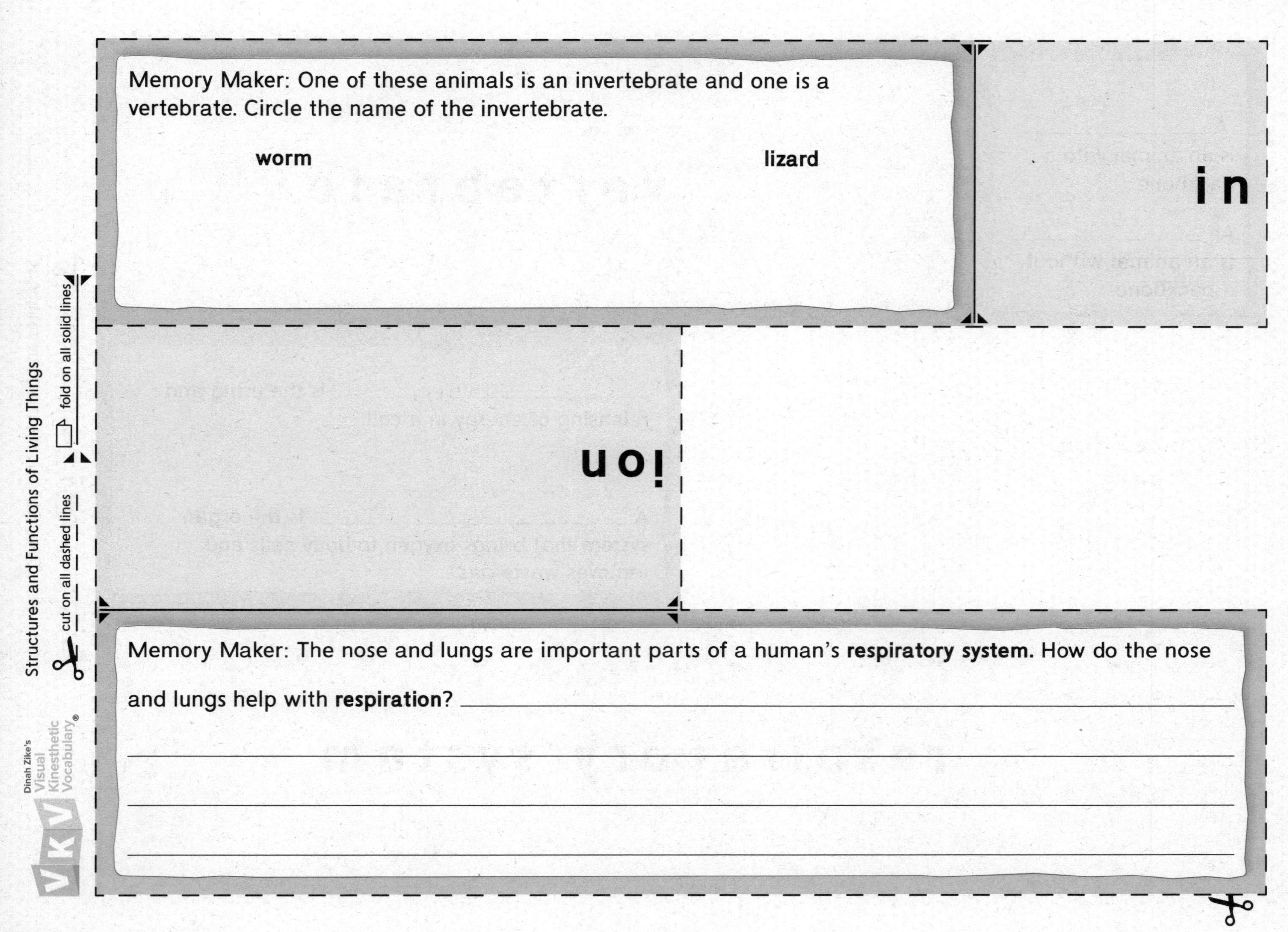

cut on all dashed lines
fold on all solid lines

Wave Patterns and Information Transfer

cut on all dashed lines

fold on all solid lines

mimicry

__________ is the act of using nature as a model for human inventions.

echo

__________ is the process of finding an object by using reflected sound.

An __________ is a repetition of a specific sound produced by reflection of sound waves from a surface.

wave

Amplitude

Wavelength

__________ is the distance from the top of one wave to the top of the next.

cut on all dashed lines
fold on all solid lines

bio

Memory Maker: A mimic copies the appearance or behavior of something else. Who is the mimic in **biomimicry**? Is it nature or the human invention? ______

______

______

location

Memory Maker: Draw a picture to define the term **echo**.

length

Memory Maker: Use your own words to explain what a **wavelength** is.

______

______

______

Wave Patterns and Information Transfer

cut on all dashed lines

fold on all solid lines

transverse wave

A ____________ is a wave that moves matter left and right as it travels through a medium.

A ____________ is a wave that moves matter up and down as it travels through a medium.

Dinah Zike's Visual Kinesthetic Vocabulary®

Wave Patterns and Information Transfer

cut on all dashed lines

fold on all solid lines

Memory Maker: How are a **longitudinal wave** and **transverse wave** alike? How are they different?

longitudinal

cut on all dashed lines

fold on all solid lines

__________ is plants that cover a particular area.

vegetation

__________ is a rock that forms when small bits of matter are pressed together in layers.

sedimentary rock

Patterns of Earth's Changing Features

cut on all dashed lines

fold on all solid lines

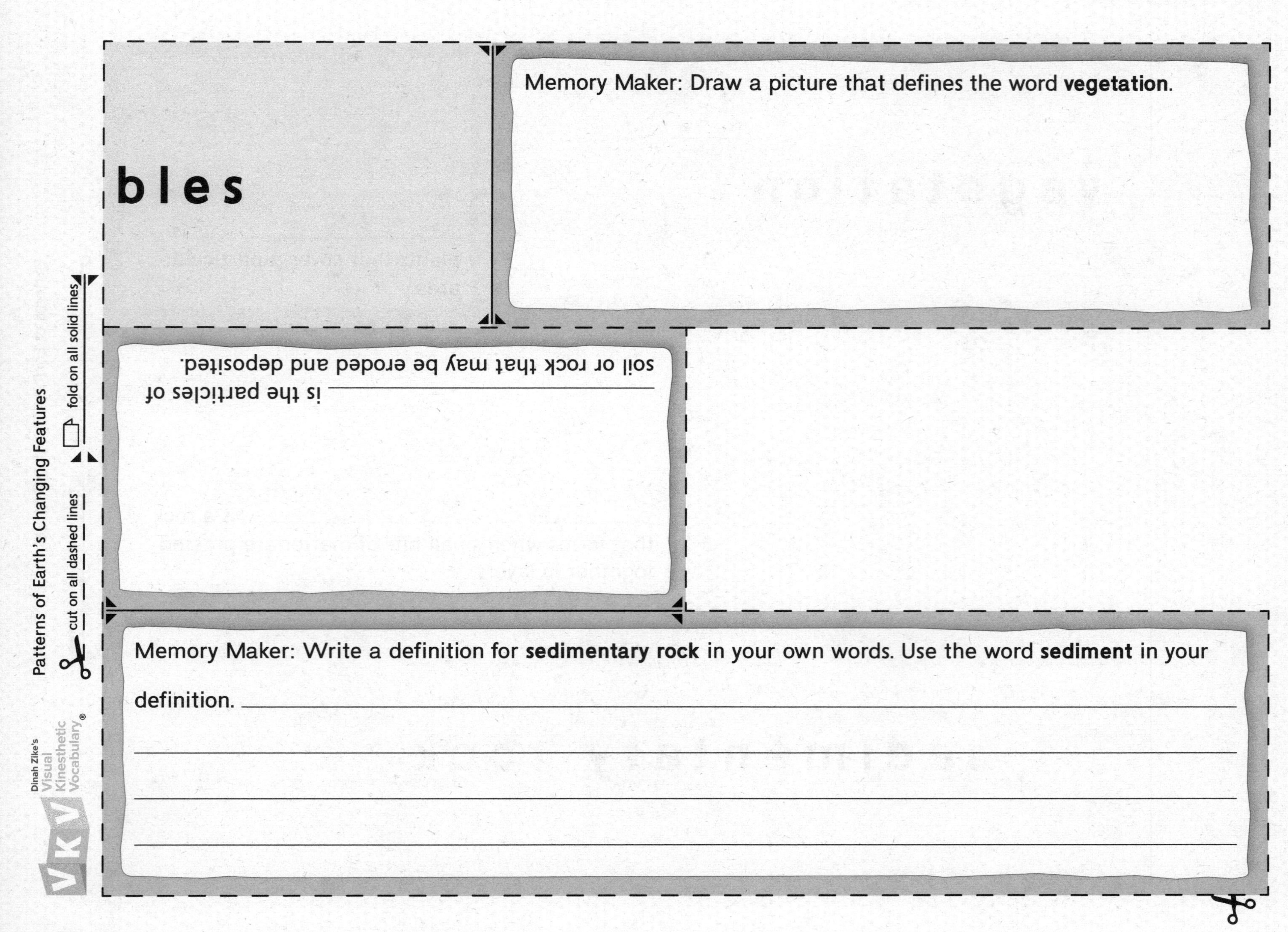

bles

Memory Maker: Draw a picture that defines the word **vegetation**.

_______ is the particles of soil or rock that may be eroded and deposited.

Memory Maker: Write a definition for **sedimentary rock** in your own words. Use the word **sediment** in your definition. _______

Patterns of Earth's Changing Features

cut on all dashed lines

fold on all solid lines

Memory Maker: Acid is a substance that is powerful enough to corrode metal. What does this tell you about the nature of acid rain? ________

____

____

____

acid

A ________ is a physical feature on Earth's surface.

A ________ is the rapid movement of rocks and soil down a hill.

landslide

cut on all dashed lines
fold on all solid lines

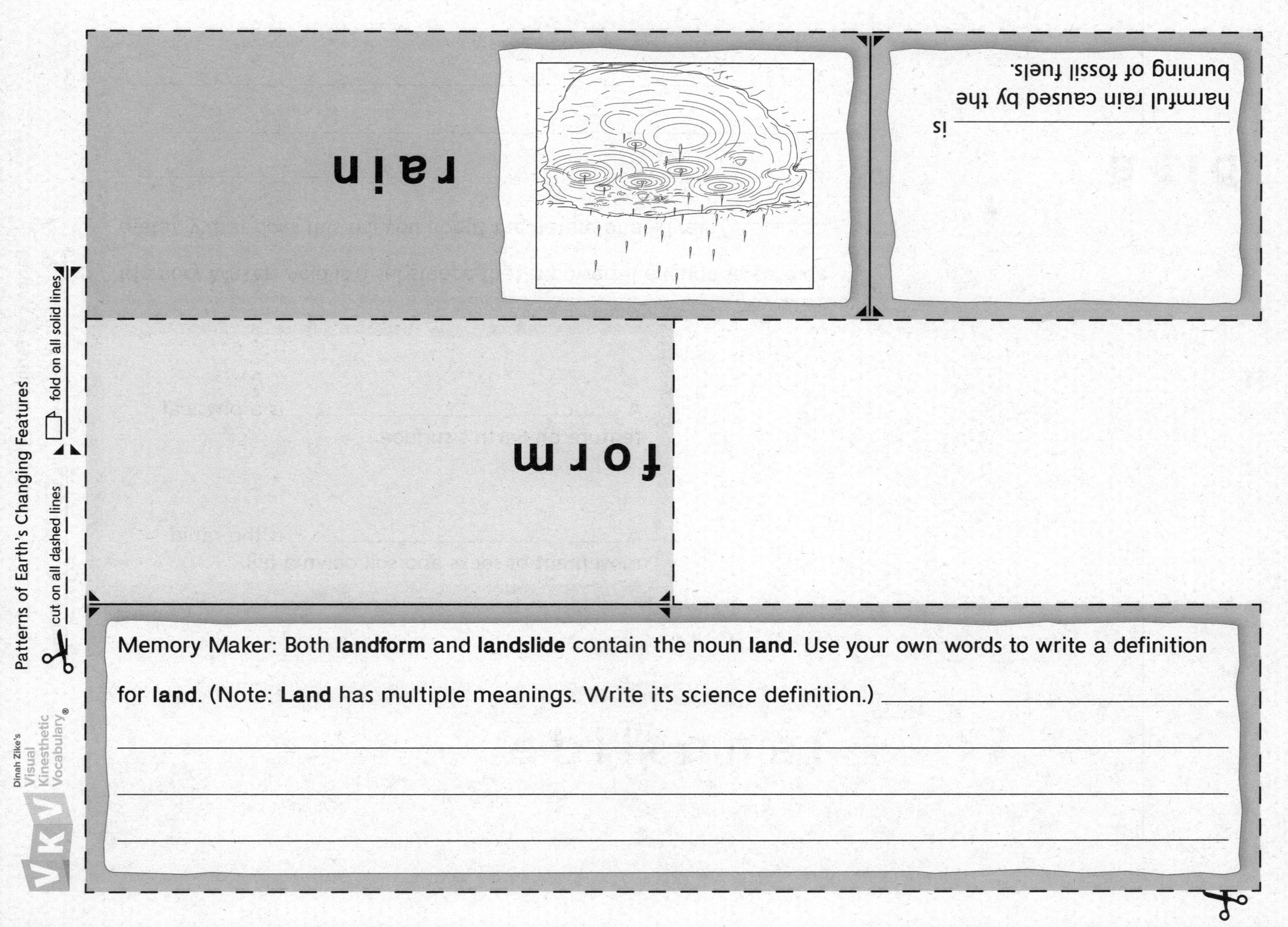

cut on all dashed lines

fold on all solid lines

tsunami

Water

A ________ is a huge wave caused by an earthquake under the ocean.

A ________ is an instrument that detects and records earthquakes; it shows seismic waves as curvy lines along a graph.

A ________ is a vibration caused by an earthquake.

seismic wave

cut on all dashed lines

fold on all solid lines

Memory Maker: Both **seismic** and **seismograph** come from the same ancient Greek word: **seismos.** What do you think **seismos** means: **record, curvy,** or **earthquake?** Why do you think that? ________

o g r a p h

Memory Maker: Construct a word web around the word **waves.** What are some different kinds of waves you have learned about?

w a v e s

cut on all dashed lines

fold on all solid lines

A ______ is something that is found in nature and is valuable to humans.

nonrenewable resource

A ______ is a natural material or source of energy that is useful to people and cannot be replaced easily.

A ______ is a useful material that is replaced quickly in nature.

cut on all dashed lines

fold on all solid lines

Memory Maker: The word part **non-** is a prefix that means "not." How does this help you remember the difference between a **renewable resource** and a **nonrenewable resource?** __________

____

____

renewable

natural